T0179640

Basic Bioscience
Laboratory Techniques

Basic Bioscience Laboratory Techniques

A Pocket Guide

Second Edition

Philip L.R. Bonner and Alan J. Hargreaves

WILEY Blackwell

This edition first published 2022
© 2022 by John Wiley & Sons Ltd

Edition History
John Wiley & Sons Ltd (1e, 2011)

The right of Philip L.R. Bonner and Alan J. Hargreaves to be identified as the authors of this work has been asserted in accordance with law.

Registered Offices
John Wiley & Sons, Inc., 111 River Street, Hoboken, NJ 07030, USA
John Wiley & Sons Ltd, The Atrium, Southern Gate, Chichester, West Sussex, PO19 8SQ, UK

Editorial Office
9600 Garsington Road, Oxford, OX4 2DQ, UK

For details of our global editorial offices, customer services, and more information about Wiley products visit us at www.wiley.com.

Wiley also publishes its books in a variety of electronic formats and by print-on-demand. Some content that appears in standard print versions of this book may not be available in other formats.

Library of Congress Cataloging-in-Publication Data

Names: Bonner, Philip L. R., author. | Hargreaves, Alan, author.
Title: Basic bioscience laboratory techniques : a pocket guide / Philip
    L.R. Bonner, Alan J. Hargreaves.
Other titles: Bioscience laboratory techniques
Description: Second edition. | Hoboken, NJ : Wiley-Blackwell, 2022. |
    Preceded by Bioscience laboratory techniques : a pocket guide / Philip
    Bonner and Alan Hargreaves. 2011. | Includes bibliographical references
    and index.
Identifiers: LCCN 2022007545 (print) | LCCN 2022007546 (ebook) | ISBN
    9781119663355 (paperback) | ISBN 9781119663447 (adobe pdf) | ISBN
    9781119663485 (epub)
Subjects: MESH: Biomedical Research | Laboratory Manual | Handbook
Classification: LCC R860 (print) | LCC R860 (ebook) | NLM W 49 | DDC
    610.72/4–dc23/eng/20220328
LC record available at https://lccn.loc.gov/2022007545
LC ebook record available at https://lccn.loc.gov/2022007546

Cover Design: Wiley
Cover Images: © Madzia71/Getty Images, Gannet77/Getty Images, Kwanchai Lerttanapunyaporn/Getty Images,
Repina Valeriya/Shutterstock.com

Set in 8.75/10.5pt TimesLTStd by Straive, Pondicherry, India

SKY10035531_072922

CONTENTS

PREFACE

In recent years, laboratory work has been severely restricted. However, knowledge of laboratory procedures is still a requirement for the successful completion of a bioscience degree. Textbooks, demonstrations, and internet videos cannot provide students with the same experience as a 'hands on' experiment. However, they can provide guidance towards a successful laboratory procedure and draw attention to safe progress through the procedures.

Following the success of the first edition of this textbook, we have been encouraged to prepare a second edition. This new edition addresses some of the changes we have seen in the laboratory classes at Nottingham Trent University. This has included increasing use of 96-well microplates for protein and enzyme assays as well as their use in immunological assays. We have included reference to the use of microplates in enzyme assays in Chapter 3 and introduced a new chapter (9) on the use of immunological assays.

The book still provides students with a reference text containing information on the basis of bioscience procedures with worked examples of the type of calculations they will encounter in their journey through a bioscience degree and beyond. The book should also provide bioscience students with a good knowledge foundation of techniques they will experience in their undergraduate and postgraduate projects.

We have retained the small book format so that students can easily transport the book and refer to the text prior to experiencing bioscience laboratory sessions.

We would like to thank our families (Alan: Olivia, Alan, Elena, Wayne, Stephanie and Tom. Phil: Elizabeth, Francesca, Mark and Neil) and friends (Wayne, Bisher, Biola, Mohamed and Ida) for their support and suggestions during the preparation of this book. We would also like to thank Rosie, Julia and Shiji at Wiley and Sons for their patience and encouragement during the book's preparation period.

Philip L.R. Bonner
Alan J. Hargreaves

GLOSSARY

Agarose: A polysaccharide with many hydroxyl groups. The material can be used to construct gels to separate nucleic acid fragments.

Ampholytes: A mixture of polycarboxylic and polyamino acids.

Antibody: A molecule that recognizes a specific antigen or antigenic epitope.

Antigen: A molecule that can be recognized by a specific antibody. Antigens are often proteins but can also include other types of macromolecules such as DNA or lipids. Different types of immunoassays exploit this property thus to detect and/or quantify antigens in cells, tissues and body fluids.

Antigenic determinant: See epitope.

Azocasein: An orange azo dye covalently bound to the milk protein casein, which provides a substrate for peptidases which cleave peptide bonds within the three-dimensional structure of a protein.

Bacilli: Rod-shaped bacteria (singular – bacillus). An example is *Bacillus megaterium*.

bis: Occurring twice, as in bis-acrylamide or bis-tris-propane.

Centrioles: Cylindrical structures that nucleate the formation of microtubules in eukaryotic cells and form the poles of the mitotic spindle.

Chelating agent: A compound (e.g. EDTA) which preferentially binds to metal ions. This reduces or effectively eliminates the metal ions' presence in a solution.

Clathrate: (caged in a lattice) A chemical structure where a compound is trapped within a lattice structure of another compound.

Cocci: Round or spherical-shaped bacteria (singular – coccus). An example is *Micrococcus luteus*.

Collimate: To make parallel.

Confidence interval: A confidence interval for a set of values describes the likely range within which a specific value can be considered to be part of the same population or data set. Increasing the degree of confidence with which this assumption can be made (e.g. from 95 to 99%) will increase the limits of the range and vice versa. This measure is often used in clinical research, as it describes the likely range of expected values for a particular treatment or condition. See also 'confidence limits'.

Confidence limits: A statistical term for the pair of values that describe the upper and lower end of a range of values within which a value can be expected to fall for a given level of confidence. In other words, the two values are the end points of the confidence interval.

Dalton: The mass of a reagent relative to 1/12 the mass of carbon, i.e. 1.0.

Dialysis: A means to separate components in a solution by unequal diffusion through a semipermeable membrane.

Dynodes: One of a series of electrodes in a photomultiplier.

ELISA: Abbreviation for 'enzyme-linked immunosorbent assay', which is an antibody-based assay in which antigen (one-site ELISA) or antibody (two-site ELISA) are typically immobilized on a solid support matrix, such as a microtiter plate. Enzyme-linked secondary antibodies are used to detect antigen–antibody binding.

Endoplasmic reticulum: A specialized membranous organelle within eukaryotic cells responsible for synthesis of membrane proteins and lipids.

Epitope: The specific part of an antigen that is recognized by an antibody. In the case of protein antigens, this could be a short sequence of consecutive amino acids in the protein antigen's primary structure. However, epitopes with chemical modifications of protein-bound amino acids (e.g. phosphorylation of tyrosine, serine and threonine residues, acetylation of lysine residues, and polyglutamylation) can also be detected by some antibodies.

Eukaryotes: Organisms that are composed of one or more cells which contain a membrane-enclosed nucleus and organelles (e.g. animal, plant, fungi, yeast and most algae).

Fab fragment: *Fragment antibody binding* can be released by limited proteolysis of an immunoglobulin molecule; it contains the antigen binding site and comprises one constant and one variable domain of each of the heavy and light chains.

Fc fragment: *Fragment crystallizable* is a fragment of an immunoglobulin molecule which can be released by limited proteolysis; it contains the 'effector region', which binds to Fc receptors on the cell surface of certain types of cells of the immune system or to proteins of the complement system thus facilitating activation of the immune system. It is the part of primary antibodies that is typically targeted by labelled secondary antibodies in immunoassays. It comprises heavy chain constant regions from both heavy chains.

Fluorophore: A molecule that emits fluorescence at a specific wavelength (emission wavelength) when irradiated with an appropriate wavelength of electromagnetic radiation (excitation wavelength). They may be used to detect the levels of specific molecules to which they bind, or they may be covalently linked to proteins such as antibodies and used in the detection of specific antigens in immunoassays.

Focal plane: The plane perpendicular to the lens' optical axis in which images of points in the object field of the lens are in focus.

Glycogen: A storage form of glucose in animal cells. Often present as cytoplasmic granules, which are particularly abundant in liver cells.

Golgi apparatus: A specialized organelle within the cytoplasm of eukaryotic cells that is involved in the glycosylation of membrane proteins and secretory proteins.

Gravity: The natural weak force of attraction that exists between all bodies in the universe ($g = 981$ cm s^{-2}).

Heavy constant (HC): A highly conserved sequence in the immunoglobulin heavy chain

Heavy variable: A sequence in the immunoglobulin heavy chain containing hypervariable regions involved in antigen binding.

Homogenate: A tissue or cellular mixture resulting from the action of a homogenizer.

Homogenizer: A laboratory machine that disrupts tissue (cells) by shearing, cutting and blending the starting material in a buffer.

Hydrophilic: ('water loving') A molecule (functional group) which prefers to interact with water or other polar solvents.

Hydrophobic: ('water hating') A molecule (functional group) which avoids contact with water or other polar solvents.

Immiscible: Incapable of mixing together (e.g. oil and water).

Immunoassays: Assays that exploit the use of antibodies to detect and/or quantify specific antigens in cells, tissues and body fluids.

Immunoglobulin: Proteins produced by circulating B cells and various types of white cells, which assist in the immune response (e.g. during infection). Different classes of immunoglobulins exist and have distinct roles in the immune systems. Antibodies are immunoglobulin molecules that can be produced and purified for use in immunoassays, whereas circulating immunoglobulins can assist in the immune response and/or represent useful biomarkers of diseases.

Intermediate filaments: Filaments of 10 nm width made of various proteins (e.g. desmin in muscle cells, nuclear lamins in the nuclear lamina). These filaments play a structural role in eukaryotic cells.

In vitro: Literally means 'in glass'.

In vivo: Literally means 'in life'.

Isoelectric point (pI): The pH value at which a zwitterionic molecule (e.g. an amino acid or protein) carries no net charge. The molecule will not move under the influence of an electrical field or bind to ion exchange resins.

I.U.: International units of enzyme activity defined as 1μmol of product formed (or substrate consumed) min^{-1} at a given temperature (usually 25 °C).

Light constant (LC): A highly conserved sequence in the immunoglobulin light chain.

Light variable (LV): A sequence in the immunoglobulin light chain containing hypervariable regions involved in antigen binding.

Lyophilization: (freeze drying) The removal of water from a solution or tissue after freezing in a vacuum, where the water sublimes from the solid phase to the gas phase without entering the liquid phase.

Lysosomes: Single membrane bounded organelles in eukaryotic cells that contain hydrolytic enzymes. These enzymes (e.g. acid phosphates) are maintained in an acidic pH environment within the organelle, where they can degrade macromolecules in a controlled manner.

Mean: The arithmetic average of a set of values.

Median: The exact midpoint of a set of values or observations.

Meniscus: The curved shape assumed by a liquid in a cylindrical tube.

Microfilaments: Polymers of 5 nm width made from the protein actin. Important in contractile processes (e.g. thin filaments in muscle cells and the formation of the contractile ring in the late stages of mitosis).

Microtubules: Tubular polymers (external diameter = 25 nm) of the protein tubulin. Microtubules play an important role in intracellular transport, cell movements and cell division, forming interphase networks and the mitotic spindle.

Mitochondria: Discrete organelles present in most eukaryotic cells involved in the production of ATP.

Mode: The most frequently observed value in a set of data.

Mr: The mass of a reagent relative to the molecular mass of hydrogen (i.e. 1.0).

Mycoplasma: A very small bacterium that lacks a cell wall. Mycoplasmas are typical contaminants of eukaryotic cell cultures.

Nomogram: A two-dimensional graphical calculating plot.

Non-parametric test: Statistical tests that do not make the assumption of a normal distribution of values and therefore can be applied to asymmetric distributions more robustly. Values are usually ranked or grouped. An example is the Mann–Whitney U test.

Normal (Gaussian) distribution: Asymmetrical or bell-shaped distribution of data values with the arithmetic mean in the exact centre.

Nucleic acids: Nucleic acids are unbranched polymers of nucleotides.

Nucleoside: The combination of a base and a pentose sugar with no phosphate groups attached. Examples (with their corresponding base in parentheses) are

- adenosine or deoxyadenosine (adenine)
- guanosine or deoxyguanosine (guanine)
- cytidine or deoxycytidine (cytosine)
- uridine (uracil) or deoxythymidine (thymine)
- Phosphorylation of nucleosides at the 5' carbon atom of the pentose sugar results in the formation of the corresponding **nucleotides**. Nucleoside triphosphates, nucleoside diphosphates and nucleoside monophosphates are nucleotides with three, two or one phosphate groups. Examples are adenosine monophosphate (AMP), adenosine diphosphate (ADP) and adenosine triphosphate (ATP).

Nucleotide: Nucleotides are made up of three components:

- A pentose sugar (ribose in RNA and deoxyribose in DNA)
- A nitrogen containing ring structure or *base*
  In DNA, the nucleotide bases are adenine (A), thymine (T), guanine (G) and cytosine (C).
  In RNA, the nucleotide bases are A, T, G and U (uracil, which replaces cytosine).
  The combination of a base and a pentose sugar is called a **nucleoside**.
- Up to three phosphate groups.
  These are attached at the 5' carbon atom of the pentose. Both DNA and RNA are assembled from nucleoside triphosphates (i.e. **nucleotides** with three phosphate groups). Adjacent nucleotides in the nucleic acid sequence are covalently linked between the 3' carbon atom of one pentose and the phosphate group attached to the 5' carbon of the next pentose. The two terminal phosphates groups are released from each nucleotide as it is added on to the polymerizing polynucleotide chain.

Nucleus: A eukaryotic membrane-bound organelle which contains the majority of the cells genetic material.

Null hypothesis: In statistical testing, the null hypothesis suggests that there is no statistical significance in a set of observations. It proposes that there is no real difference between two variables or one variable and zero that could not have arisen by chance. For the null hypothesis to be rejected, an appropriate statistical test must be performed producing a level of confidence (typically $p < 0.05$) indicating that the differences are significant.

Oil immersion: This is a technique that involves placing a drop of immersion oil, which is transparent and has a high refractive index, between the ×100 objective lens and the cover slip covering the specimen. The image is focussed with the lens in contact with the oil droplet, which greatly enhances the resolution.

Oxidation: The removal of electrons.

Parametric test: These are statistical tests that make the assumptions that the values in each population or data set are normally distributed and that they have equal variances. An example is the student's t-test. If there are significant deviations from these assumptions, a non-parametric test should be used.

Paratope: The region of an antibody that recognizes a specific antigenic determinant (epitope).

Peroxisomes: Small single membrane bounded organelles in eukaryotic cells that contain enzymes involved in fatty acid metabolism and breakdown of peroxides. The latter is catalysed by the enzyme *catalase*, which is enriched in peroxisomes and is used as a biochemical marker enzyme to monitor peroxisome enrichment in cell fractionation procedures.

Peptidoglycan: A covalently cross-linked network of peptides and polysaccharides that make up the structure of bacterial cell walls. It is also referred to as *murein*.

pI: The isoelectric point is the point on a pH scale where the positive charges on a molecule are balanced by the negative charges, i.e. no net charge.

pka: The negative $\log_{10}$ of the equilibrium constant (Ka), i.e. $pKa = -\log_{10} Ka$. The acid and basic forms are in equal concentration.

Plasma membrane: A phospholipid and protein fluid bilayer surrounding cells.

Primary antibody: An antibody that is directed against the specific antigen of interest in a given immunoassay.

P value: The probability that an observed value or mean value could have arisen by chance alone. For example, if two sets of values are compared using an appropriate statistical test and the resultant p value is less than 0.05, there is less than a 5% chance that the differences could have occurred by chance. Therefore, the null hypothesis of 'no real difference' is rejected and the two sets of data are considered to be significantly different.

Reduction: The addition of electrons.

Reducing agents: A substance (e.g. DTT or 2-ME) that chemically reduces other substances (adding electrons).

Resolution: Base line separation of two compounds in chromatography.

Resolving gel: A polyacrylamide gel used to separate complex protein mixture.

Ribosomes: Macromolecular structures involved in the synthesis of proteins. They can exist as membrane-bound ribosomes (e.g. in the rough endoplasmic reticulum) or 'free' polyribosomes.

Secondary antibody: A labelled antibody used to detect the binding of a primary antibody to its corresponding antigen.

Spatula: A small implement with a flat blade that can be used to help weigh and mix reagents.

Stacking gel: a polyacrylamide gel which is layered over the resolving gel.

TARE: Unladen weight (e.g. the weight of a beaker on a balance). By subtracting the TARE weight from the gross weight, the weight of the reagent can be provided.

T value: The t value is the value obtained when a t test is performed to compare two sets of data. It measures the difference between the means of the two sets or populations of data taking into account the variation between values. Data sets used in such analyses must exhibit a normal distribution.

van der Waals: A weak intermolecular force.

Variance: A measure of dispersion that is equivalent to the mean of the squares of the difference of each value from the population mean. It is the square of the standard deviation.

Zwitterionic: A molecule with both acidic and basic residues (e.g. amino acids or proteins).

ABBREVIATIONS

A_{260nm}:	Absorbance at 260nm
ΔA_{340nm}:	The change in absorbance at 340nm
A:	Adenine
AMPS:	Ammonium persulphate
AP:	Alkaline phosphatase
ATCC:	American Type Culture Collection
ATP:	Adenosine 5'-triphosphate
BApNA:	Benzoyl-L-arginine para-nitroanalide (this is an older version of chemical nomenclature, which has been superseded by benzoyl-L-arginine 4-nitroanalide)
Bar:	The metric unit of pressure
Bp:	Base pair
BSA:	Bovine serum albumin
C:	Cytosine
C5:	A chain of 5 carbon atoms
C8:	A chain of 8 carbon atoms
C18:	A chain of 18 carbon atoms
CDR:	Complementary determining region
CFSE:	Carboxy fluorescein succinimidyl ester
CFU:	Colony forming unit
CHAPS:	(3-[(3-Cholamidopropyl) dimethylammonio]-1-propane sulphonate
Da:	Dalton
DAPI:	4'6-Diamidino-2-phenylindole
ddA:	Dideoxyadenosine triphosphate
ddC:	Dideoxycytidine triphosphate
ddG:	Dideoxyguanosine triphosphate
ddT:	Dideoxythymidine triphosphate
DF:	Dye front
DIC:	Differential interference contrast
DMSO:	Dimethyl sulfoxide
DNA:	Deoxyribonucleic acid
DTT:	Dithiothreitol
(ε_{340nm}):	The molar absorptivity coefficient to be used in the Beer Lambert Law.
$(\varepsilon^{1\% \, (w/v)})$:	The mass absorptivity coefficient for protein solutions at a given wavelength (e.g. 280nm) to be used in the Beer Lambert Law
ECL:	Enhanced chemiluminescence
EDTA:	Ethylenediaminetetraacetic acid
ELISA:	Enzyme-linked immunosorbent assay
Fab:	Antibody binding fragment (of an Ig)
Fc:	Cell binding fragment
FITC:	Fluorescein isothiocyanate
$FMNH_2$:	Flavin mononucleotide (reduced)
FR:	Framework regions
G:	Guanine
GAPDH:	Glyceraldehyde-3-phosphate dehydrogenase
GLC:	Gas liquid chromatography
HC:	Heavy constant (domain)
H & E:	Hematoxylin and eosin

HEPES:	N-2-hydroxyethylpiperazine-N'-2-ethanesulphonic acid
HETP:	Height equivalent to theoretical plate
HIC:	Hydrophobic interaction chromatography
HiLiC:	Hydrophilic interaction liquid chromatography
HPLC:	High pressure liquid chromatography
HRP:	Horseradish peroxidase
HV:	Heavy variable region e.g. of an antibody
IEX:	Ion exchange chromatography
IPG:	Immobilised pH gradient
I.U.:	International units of enzyme activity
Ka:	The equilibrium constant (association/dissociation).
Kav:	An approximation to K_D in size exclusion chromatography
kDa:	kilo Daltons (1 x 10^3 Daltons)
K_D:	The equilibrium (distribution or partition) coefficient in chromatography or the dissociation equilibrium constant (antibodies)
LC:	Light constant
LV:	Light variable
2-ME:	2-Mercaptoethanol
MeOH:	Methanol
MES:	2(N-morpholine)ethane sulphonic acid
Mr:	Relative molecular mass
MTT:	(3[4,5-dimethylthiazol-2-yl]-2,5 diphenyl tetrazolium bromide
MW:	Molecular weight
NAD^+:	Nicotinamide adenine dinucleotide (oxidised)
NADH:	Nicotinamide adenine dinucleotide (reduced)
NBT:	Nitro blue tetrazolium
NMR:	Nuclear magnetic resonance
NP:	Normal phase in chromatography
OD:	Optical density (used in microbiology)
ODS:	Octadecyl silane (C18) a widely used chromatography ligand
PAGE:	Polyacrylamide gel electrophoresis
2D-PAGE:	Two-dimensional polyacrylamide gel electrophoresis
PBS:	Phosphate-buffered saline
PCR:	Polymerase chain reaction
PEEK:	Polyether ether ketone
PES:	Polyether sulphone
PFGE:	Pulsed field gel electrophoresis
PFU:	Plaque forming unit
PIPES:	1,4-Piperazine-bis (ethane sulfonic acid)
PMS:	Phenazine methosulfate
PSI:	Pounds per square inch (Imperial measurement of pressure)
PVDF:	Polyvinylidene difluoride
RCF:	Relative centrifugal force
RER:	Rough endoplasmic reticulum
Rf:	Retention factor (relative mobility or relative to the front)
RIA:	Radioimmunoassay
RNA:	Ribonucleic acid
RP:	Reversed phase
RPM:	Revolutions per minute
RT-PCR:	Reverse transcriptase polymerase chain reaction
S:	Standard deviation (in statistical formulae)
S^2:	Variance
SAX:	Strong anion exchange

SCX:	Strong cation exchange
SD:	Standard deviation
SDS:	Sodium dodecyl sulphate
SDS-PAGE:	Polyacrylamide gel electrophoresis in the presence of SDS
SEC:	Size exclusion chromatography
SEM:	Standard error of the mean (in statistical analysis) or scanning electron microscopy
SER:	Smooth endoplasmic reticulum
SPE:	"Solid phase extraction" or "sample preparation and extraction" in chromatography
Std:	Standard
T:	Thymine
TAE:	Tris-acetate-EDTA
TARE:	The weight of the container on a balance pan
TBE:	Tris-borate-EDTA
TEM:	Transmission electron microscopy
TEMED:	N,N,N', N'- tetramethylethylenediamine
TCA:	Trichloroacetic acid
TLC:	Thin layer chromatography
TMB:	Tetramethyl benzidine
Tricine:	N-[tris(hydroxymethyl)methyl] glycine
Tris:	2-Amino-2-hydroxymethylpropane-1,3-diol.
TRITC:	Tetramethyl rhodamine isothyocyanate
uHPLC:	Ultra-high pressure liquid chromatography
UV:	Ultraviolet light
Ve:	The elution volume of a sample applied to a size exclusion column
Vo:	The void volume of a size exclusion column
Vt:	The total volume of a size exclusion column
(v/v):	Volume in a volume
WAX:	Weak anion exchange
(w/v):	Weight in a volume

ABOUT THE COMPANION WEBSITE

This book is accompanied by a companion website:

www.wiley.com/go/bonner/basiclab2

This website includes:

- PowerPoints of all figures
- PDF of all tables

1

THE PREPARATION OF SOLUTIONS
IN BIOSCIENCE RESEARCH

1.1 Introduction

To help understand the many complex processes and interactions that occur within a cell (in vivo), it is often necessary to simulate cellular events in the test tube (in vitro). The many constituents of the cell vary in their concentration, size and physical characteristics, and while it is often difficult to exactly simulate the cellular environment in the laboratory, it is possible to provide an environment in which the cellular constituent under study functions normally. Then, by careful adjustment of the solution, an understanding of the cellular constituent's role within the cell can be ascertained. This means that the preparation of solutions is a vital step in bioscience research and a topic that all bioscience students should become familiar with.

However, to gain complete control of this topic, you *must* memorize key topics, such as the common units used in bioscience research (Table 1.1). The only way to avoid confusion is to commit to memory the various units involved or to keep a copy of Table 1.1 conveniently placed for ease of use. The units detailed in Table 1.1 are representative of the concentrations at which many metabolites are present within the cell, for example, calcium ions (Ca^{2+}) can be present at mM concentrations outside the cell but only nM within the cell's cytoplasm.

Student Exercise

- Memorize the units in Table 1.1
- Memorize the sub-divisions to the units in Table 1.1.

Having mastered the units, you will be ready to apply the information to the preparation of solutions, although one of the major stumbling blocks you may find is overcoming an aversion to calculations. Many students view biosciences as descriptive subjects and try to block out the numerical aspects. If this is the case for you, then this approach will greatly limit your involvement in the topic. In fact, this fear of numeracy is usually misplaced as most of the calculations are straightforward and can become second nature with practice.

On a more practical note, from a student's point of view, numerical assessments within bioscience courses differ from descriptive topics in one notable aspect. In an assessment, it is possible to get 100% of the available marks on a numerical question, whereas this is rarely true for a descriptive answer, as the marker will always be reluctant to allocate 100% of the marks, signifying a perfect answer.

Basic Bioscience Laboratory Techniques: A Pocket Guide, Second Edition. Philip L.R. Bonner and Alan J. Hargreaves.
© 2022 John Wiley & Sons Ltd. Published 2022 by John Wiley & Sons Ltd.

Table 1.1 Common units in bioscience.

	1.0	1.0×10^{-3}	1.0×10^{-6}	1.0×10^{-9}	1.0×10^{-12}
Concentration	Molar (M)	millimolar (mM)	micromolar (μM)	nanomolar (nM)	picomolar (pM)
Amount	Mole (mol)	millimole (mmol)	micromole (μmol)	nanomole (nmol)	picomole (pmol)
	gram (g)	milligram (mg)	microgram (μg)	nanogram (ng)	picogram (pg)
Volume	Litre (l)	millilitre (ml)	microlitre (μl)	nanolitre (nl)	picolitre (pl)

1.2 Concentration

The concentration of metabolites, chemicals and reagents that are used in bioscience is reported in a few different notations (e.g. molarity and percentage).

1.2.1 Molarity

Science uses molarity to ensure numerical equivalence in terms of the molecules present in a solution. In one mole (symbol: mol) of any chemical, there are 6.022×10^{23} molecules present (Avogadro's number). This is true for all reagents, for example, one mol of sodium chloride (NaCl; Mr 54.55), glucose ($C_6H_{12}O_6$; Mr 180.16) or adenosine triphosphate di sodium hydrate (ATP; Mr 551.14) will all have the same number of molecules present. One mol of these reagents will result in different weights because different compounds will have different atoms present in their structure, in effect a different size or mass. But one mol of a reagent will *always* have the same number of molecules present.

Molarity (M) is the most common method of describing the concentration of a reagent in solution. The sentence should be read again 'the *concentration* of a reagent *in solution*'. This describes exactly what occurs in that a reagent (possibly a powder) is dissolved in a volume of liquid (usually water). Therefore, two things are required before a concentration can be ascribed; a known weight of a reagent must be physically weighed out, then transferred and dissolved in a known volume of liquid. One mol of a compound is a physical weight (relative molecular mass (Mr) in grams). You can go to the chemical shelves, take the reagent required and weigh out one mol.

When a reagent is dissolved in a volume of liquid, the units change from mol to molarity (M; a unit of concentration). The shorthand version of this unit is the capital letter M (Table 1.1). This unit of concentration describes an amount of reagent dissolved in a volume of liquid.

A molar solution (M) is the molecular weight in grams (1 mol) dissolved in a litre (1000 ml) of water (see Table 1.1).

$$A\,1.0\,M\,solution = \frac{\text{The molecular weight in grams of a reagent}\,(1.0\,\text{mole})}{\text{Volume}\,(1.0\,l\,\text{or}\,1000\,ml)}$$

1.3 Using a Balance to Weigh Out Reagents

1.3.1 The Use of an Electronic Balance

In practice, the amount of reagent you would weigh out to prepare a solution depends on the availability of the reagent (always check, sign and comply with the Health and Safety data sheet for the reagent before commencing any weighing out) and access to an appropriate accurate balance.

Most bioscience laboratories will have a range of balances accurate to 1, 2, 3 or 4 decimal places. Large amounts of inexpensive reagents are routinely weighed out on 'top pan' balances (see Figure 1.1), whereas small amounts of expensive reagents should be weighed out using an analytical balance. If a hazardous reagent is being weighed out, use a balance contained within a designated safety area which will have a negative air pressure to ensure that no particles of the reagent can enter and contaminate the laboratory air space.

(a)

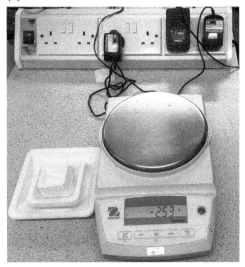

(b)

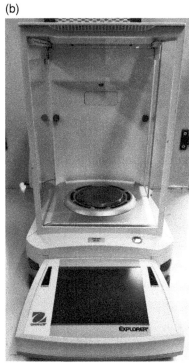

Figure 1.1 Examples of typical balances (a) top pan (b) analytical.

Most if not all balances will have both a 'Zero' button and a 'TARE' button. Before using a balance, close all the access doors and press the 'Zero' button. This sets the readout to zero without any weight on the balance pan. Place a reagent container on the balance, close the access doors and set the readout to zero again, this time using the 'TARE' button. The TARE takes into account the weight of the reagent holder and resets the balance to zero, allowing the user to accurately weigh out only the mass of the required reagent.

1.3.2 The Use of a 'Top Pan' Balance

Always check to see that the balance is clean. If you need to clean the balance, take the appropriate health and safety precautions and dispose of the waste in a prescribed container.

- Press 'Zero' to set the readout on the balance to zero.
- Place a suitable container (e.g. beaker, plastic weighing boat or aluminium foil) gently onto the top pan balance (see Figure 1.1a) and wait for the electronic readout to stabilize.
- Press the 'TARE' button on the balance to bring the reading to zero.
- Using an appropriate spatula (see Figure 1.2a), gently add the reagent to the container to achieve the correct weight required.
- After you have taken the correctly weighed reagent from the balance, check that there is no spillage on the balance pan or in the surrounding areas. This is important to prevent cross-contamination of other people's solutions and to prevent health and safety issues for the next person who comes to use the balance.

1.3.3 The Use of Analytical Balances

Analytical balances are sensitive to surrounding vibrations and are usually placed on a vibration-free bench (see Figure 1.1b).

- Always check that the balance is clean, if it requires cleaning, take the appropriate health and safety precautions (see above).
- Close the access doors and press the 'ZERO' button to set the balance to a readout of zero.
- Open the side door to the balance and place a suitable container on the balance pan (e.g. small beaker, a plastic tube, plastic weighing boat or aluminium foil).
- Close the door and allow the electronic readout to stabilize. Press the 'TARE' button on the balance to bring the reading to zero. Alternatively, note the weight of the beaker and add the weight of the beaker to the amount of reagent required.
- Add the reagent to the container on the pan of the analytical balance using a micro spatula (see Figure 1.2b), close the side door and allow the read out to stabilize. To weigh out accurately is a skill, but with practice it is possible to weigh out as little as 1.0 mg of a reagent.
- When the desired amount of reagent has been weighed out, the container should be carefully removed and the reagent dissolved.
- Remember to check that the floor and the pan of the balance are clean. Usually, there are brushes and wipes available to keep these important laboratory devices clean (remember to dispose of any residues in compliance with health and safety regulations).

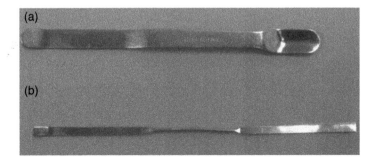

Figure 1.2 *A typical (a) spatula and (b) micro spatula.*

1.4 Practical Considerations When Making a 1.0 M Solution

No matter how careful you are in preparing a solution for use in bioscience research, there is always the possibility that bacteria or fungi may be present initially in small numbers in the final solution. This may not be a problem if the solution is to be used immediately, but it can be a problem if the solution is left in storage for a period of time.

- Storing solutions at a reduced temperature will arrest or slow down bacterial growth. A solution that is in regular use should be stored in a refrigerator (4–8 °C) and then warmed to the required operating temperature before use. Stocks of solutions can be stored frozen (25 °C) in convenient aliquots, these can then be thawed and warmed to the required operating temperature before use.
- A concentrated stock solution (e.g. ×10) may help prevent (or arrest) the growth of bacteria or fungi. Dilutions can be prepared from the stock solution (see Section 1.5), which will take up less storage space.
- If the reagent is heat stable, the prepared solution can be placed in a suitable container and heated to 120 °C under pressure in an autoclave. Allow the solution to cool before storage.
- Alternatively, before storage, the solution can be passed through a filter with a pore size (0.2 or 0.45 μm pore diameter) which is small enough to trap small particles and bacteria. Many bacteria have a diameter of approximately 0.5 μm (500 nm).
- Another approach involves using aseptic techniques to prepare the solution in a filtered air stream (see Chapter 8 for further details).
- Different combinations of the above may be required.

Working out the amount required to make one litre of a 1.0 M solution of any reagent is relatively straightforward. You can find the relative molecular mass (Mr) of the reagent from the reagent bottle, or from a reference source, accurately weigh the required amount, transfer the solid to a suitable beaker (or vessel) and then add the highest quality water available[1] initially to a volume less than 1000 ml to dissolve the reagent (see Worked Example 1.1).

Dissolving a reagent completely is necessary for the correct final concentration to be achieved. It is usual to add a magnetic stirring bar ('flea') into the solution in a beaker and then place this on a magnetic stirring device. Set the device to a slow stir and wait for the reagent to dissolve (see Figure 1.7). Reagents that are difficult to dissolve may require the addition of heat, for which hot plate stirrers can be used in conjunction with a glass beaker (remember to use the appropriate safety precautions). Alternatively, a microwave oven can be used to slightly elevate the temperature of the liquid, making sure that the container is microwaveable. In either case, boiling the water with the reagent should be avoided. Ultrasonic baths can also be used to quickly disperse solids into a solution but remember that prolonged use of these baths will eventually elevate the temperature of the solution. An increase in temperature can be moderated by the inclusion of ice in the liquid of the ultrasonic bath.

If the pH of the reagent requires (see Section 1.7.1) adjustment, the solution should be first cooled to the temperature at which it is to be used (the reason for this is that some reagents have a different pH at different temperatures, e.g. Tris/HCl buffer). The pH can be adjusted (see Figure 1.7) by the addition of an appropriate acid or base before the solution is added to a volumetric flask (see Figure 1.3a).

Worked Example 1.1 Prepare a 1.0 M solution of NaCl

- A 1.0 M solution of NaCl will have 54.5 g of NaCl in 1000 ml.
- Weigh out 54.5 g of NaCl into a 1000 ml beaker on a top pan balance.
- Add 750 ml of distilled water to dissolve the NaCl. This can be helped by the addition of a stirring 'flea' and then placing the beaker on a magnetic stirring device.
- When the solid has dissolved, transfer the liquid to a 1000 ml volumetric flask. Volumetric flasks have a mark etched on the neck of the vessel, so that when the meniscus of the liquid touches this mark the correct volume is achieved.

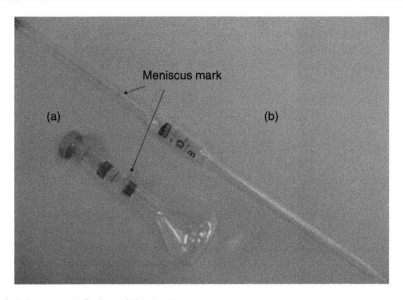

Figure 1.3 *Volumetric (a) flasks and (b) pipettes.*

- Add distilled water to the flask until the meniscus of the liquid is at the mark on the neck of the volumetric flask.
- Add a stopper and invert the flask 5–10 times to mix the liquid.
- Transfer the liquid in the volumetric flask into a storage bottle/flask. Label this with (i) the name of the reagent, (ii) the molarity, (iii) the pH, (iv) any specific health and safety precautions (e.g. corrosive or poisonous), (v) the date of preparation and (vi) your own name and course.
- Store the liquid in the fridge, freezer or at room temperature.

1.4.1 Preparing Solutions with a Concentration Less than 1.0 M and a Volume Less than a Litre

As the concentrations of metabolites within the cell are rarely in the molar range, less concentrated solutions are often required both for convenience and to reduce costs. It is these calculations that open a variety of errors for many students, but if you break the calculation down to its constituent parts and practice, the calculation will become second nature (see Worked Example 1.2).

Worked Example 1.2 Prepare 25 ml solution of adenosine triphosphate (ATP) at a final concentration of 50 mM

- *Write down what weight of ATP would be required for a 1.0 M solution*: A mass of 551.14 g of ATP disodium salt hydrate must be weighed out and dissolved in 1.0 l (1000 ml) to make a 1.0 M solution. Clearly, with expensive reagents this is a prohibitive amount.
- *Adjust for the volume required.*

We do not want 1000 ml (1.0 l), as we only require 25 ml (1.0 ml is 10^{-3} of 1.0 l). So, divide the Mr of ATP by 1000 to get to 1.0 ml and then multiply by 25 to get to the required volume (25 ml). This is now the weight in grams of ATP required to make 25 ml of a 1.0 M solution.

$$\frac{551.14 \times 25}{1000} = 13.78\,g\ldots(A)$$

You are preparing a solution that requires a smaller volume than a litre, so always check that the weight calculated is less than the Mr of the required reagent. *Adjust for the concentration.*

We do not want 25 ml of a 1.0 M solution; we require a final concentration of 50 mM (1.0 mM is 1.0×10^{-3} of 1.0 M). So, divide the weight of ATP obtained above (A) by 1000 to get to 1.0 mM and then multiply by 50 to get to the required molarity.

$$\frac{13.78 \times 50}{1000} = 0.69\,g\ldots(B)$$

This is the weight (0.69) of ATP required to make 25 ml of a 50 mM solution.

This may seem like a basic piece of advice for graduate students, but everyone can make a mistake in these calculations, and if you stick to this method you should avoid making too many errors.

There are sometimes requirements for even less concentrated solutions. The series of calculations outlined above can also be used for these less concentrated solutions.

Worked Example 1.3 Prepare 50 ml solution of ATP at a final concentration of 750 µM

- *Write down what weight of ATP would be required for a 1.0 M solution* A mass of 551.14 g of ATP has to be weighed out and dissolved in 1.0 l (1000 ml) to make a 1.0 M solution.
- *Adjust for the volume required.*

We do not need to make 1000 ml (1.0 l), as we only require 50 ml (1.0 ml is 1.0×10^{-3} of 1.0 l). So, divide the Mr of ATP by 1000 to get to 1.0 ml and then multiply by 50 to get to the required volume. This is now the weight in grams of ATP required to make 50 ml of a 1.0 M solution.

$$\frac{551.14 \times 50}{1000} = 27.56\,g\ldots(A)$$

If, as in the example above, you are preparing a solution that requires a smaller volume than a litre, always check that the weight calculated is less than the Mr of the required reagent *Adjust for the concentration.*

You do not want 50 ml of a 1.0 M solution; we require a final concentration of 750 µM (1.0 µM is 1.0×10^{-6} of 1.0 M). So, divide the weight of ATP obtained above (A) by 1 000 000 to get to 1.0 µM and then multiply by 750 to calculate the weight needed for the required molarity.

$$\frac{27.56 \times 750}{1000 \times 1000} = 0.02\,g\ldots(B)$$

This is the weight (0.02 g) of ATP required to make 50 ml of a 750 µM solution.

Using this method, it is possible to correctly calculate the amount in grams of any reagent required to make up any concentration or volume.

1.4.2 Preparing Molar Solutions of Liquids

In a similar way to solids, molar solutions of liquids can be prepared by weighing the required amount of liquid in grams into a beaker on a balance, before the addition of water. For example, to prepare 1.0 M ethanol (Mr: 46.07), a beaker is placed onto a balance, and the ethanol is pipetted into the beaker until the reading shows 46.07 g. The beaker can be removed and water added to make the volume to 1000 ml. The solution of ethanol is now 1.0 M. This is not the same as taking 46.0 ml of ethanol and making the volume to 1000 ml to prepare a 1.0 M solution of ethanol because ethanol has a different density to water. Liquids have a density relative to the density of water (1.0 g ml^{-1}); therefore, liquids with a lower density than water include ethanol (0.78 g ml^{-1}) and liquids with a higher density than water include glycerol (1.26 g ml^{-1}).

Density (ρ) is defined as the space between molecules and can be calculated by dividing the mass (kg or g) by the volume (1000 ml). Density is not related to viscosity (resistance to flow), but an increase in temperature will increase the space between molecules allowing a liquid to flow with less resistance.

$$\text{Density}\left(\rho\right)=\frac{\text{weight}\left(\text{g}\right)}{\text{volume}\left(\text{ml}\right)}$$

Thus, the volume of ethanol required to prepare a 1.0 M solution without using a balance and taking into account the density of ethanol (0.78g ml^{-1}) is calculated as follows:

$$\text{volume required}\left(\text{ml}\right)=\frac{\text{Weight required}\left(\text{g}\right)}{\text{density}\left(\text{g ml}^{-1}\right)}\quad\frac{46.07\text{ g}}{0.78\text{ g}}$$

$$= 59.0 \text{ ml of ethanol into a volumetric flask and}$$
$$\text{the volume made to 1000 ml with water}$$
$$= 1.0 \text{ M solution of ethanol.}$$

In general, solvents are routinely used as percentage solutions (see below).

1.4.3 Preparing Percentage Solutions

In both chemistry and biosciences, percentage volume/volume (v/v) solutions and weight/volume (w/v) solutions are routinely used particularly in chromatography (see Chapter 7) and cell culture (see Chapter 8).

A percentage solution means that an amount of reagent in grams is dissolved for every 100 ml of liquid. When the percentage unit is (w/v) it is a weight of a reagent in a 100 ml volume (w/v) (e.g. 0.9% w/v NaCl used in phosphate buffered saline (abbreviated to PBS).

This means that if 100 ml of this solution is required, it is prepared by weighing out 0.9 g of NaCl into a suitable vessel. A volume of water less than 100 ml is added and the NaCl is dissolved. When the NaCl has dissolved, the solution can then be made up to 100 ml to give the final percentage solution of 0.9% (w/v) NaCl.

If a different final volume is required, then both the amount of reagent and volume should be adjusted accordingly (see Worked Example 1.4).

Worked Example 1.4 The Preparation of % Solutions

- Prepare 25 ml of 0.9% (w/v) NaCl
Adjust for the volume by dividing by 100 to get the amount required for 1.0 ml and then multiply by 25 to get the amount required for 25 ml.

$$= \frac{0.9 \times 25}{100} g$$

$$= 0.225 g$$

Dissolve this amount of reagent in a volume less than 25.0 ml; when the NaCl has dissolved, make the volume up to 25.0 ml.

- Prepare 375 ml of 0.9% (w/v) NaCl

Adjust for the volume by dividing by 100 to get the amount required for 1.0 ml and then multiply by 375 to get the amount required for 375 ml.

$$= \frac{0.9 \times 375}{100} g$$

$$= 3.375 g$$

Dissolve this amount of NaCl in a volume less than 375.0 ml; when the reagent has dissolved, make the volume up to 375.0 ml.

Liquids can also be weighed out to give a weight in a volume solution, for example, 2% (w/v) glycerol. If 100 ml of 2% (w/v) glycerol were required, 2.0 g of the liquid glycerol would be weighed into a beaker and water added to a volume less than 100 ml. When the glycerol had dissolved, the volume could then be made up to 100 ml.

However, liquids are frequently prepared as a volume within a volume concentration (v/v) (e.g. 5% (v/v) methanol used in chromatography – see Chapter 7). If 100 ml of 10% (v/v) methanol solution were required, 10.0 ml of methanol would be measured out and made up to 100 ml with distilled water. Smaller volumes of the same percentage can be prepared by adjusting the volumes of both liquids proportionally.

1.4.3.1   Weight in a Volume (w/v) Solutions

Some common materials encountered in bioscience, for example proteins and nucleic acids, are commonly dissolved as a weight in a declared volume, for example, $1.0\,mg\,ml^{-1}$ (this can also be written as mg/ml). The reason for this is that solutions of proteins can be a heterogeneous population of different molecules, which makes constructing molar concentrations meaningless. Also, the concentrations of proteins from cellular extracts are commonly estimated using reagents that interact with proteins to produce a coloured product (see Chapter 3, Section 3.8).

1.5   Dilutions and the Use of Pipettes

It was proposed (Section 1.4) that a convenient method to prepare solutions of different concentrations is to initially prepare a concentrated stock solution and then dilute down from this concentrated solution to the concentration and volume required. The ability to prepare accurate dilutions is an important skill in bioscience research (see also Chapters 2 and 8).

Preparing large volume dilutions can be achieved using a measuring cylinder (see Figure 1.4b), but accurate dilutions require the use of a calibrated pipette. Volumetric glass (see Figure 1.3b) or cylindrical graduated glass/plastic pipettes (see Figure 1.4a) are still in use, but these have been superseded by the use of mechanical (or electronic) micropipettes (see Figure 1.5), primarily on health and safety grounds, accuracy and ease of use. All pipettes work by creating a vacuum above the liquid that requires dispensing in a liquid holding barrel. The liquid is dispensed by releasing the vacuum with care.

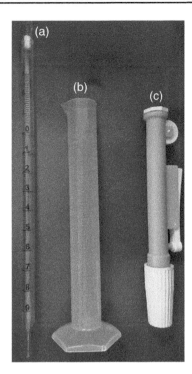

Figure 1.4 *An example of (a) a graduated pipette, (b) a measuring cylinder and (c) a pipette filler.*

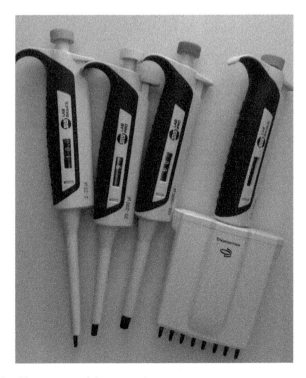

Figure 1.5 *Example of four common laboratory pipettes.*

KEY POINTS TO REMEMBER 1.1

- A dilution implies that the concentration of the final working solution is *less* than the concentration of the stock solution. If your calculation produces a value which has a higher concentration than the stock solution, you must start the calculation again. It is also important to ensure that the units of volume and concentration (see Table 1.1) are the same on both sides of the equation.
- A 1 in 10 dilution (1:10) is 1 part in a final volume of 10 (*not* 1 part added to 10 parts).

1.5.1 The Use of Graduated Cylindrical Glass or Plastic Pipettes

If a graduated cylindrical glass or plastic pipette is required to prepare a dilution from a stock solution:

- *Never* use your mouth to draw liquid into the pipette!!
- Ensure that the pipette is clean and has no flaws at either end.
- *Hold the pipette at the top* and insert it into the nozzle of a pipette filler (see Figure 1.4c or Figure 8.1) using a firm hold. Do not try to force the pipette into the filler; if there is an obviously poor grip between the filler and the pipette, exchange the pipette filler for a new one.
- Draw liquid into the cylindrical barrel to the level required, as recommended by the manufacturer. Hold the pipette above the stock liquid and check to see if the pipette drips liquid. If the pipette drips liquid, replace the pipette (or the pipette filler/or both).
- Dispense the stock solution into a suitable vessel.
- Place the used pipette in a suitable tray for cleaning.

1.5.2 The Use of Handheld Micropipettes

Micropipettes are routinely used in bioscience laboratories and, although there are many different manufacturers of handheld micropipettes, the final design is inherently similar (see Figure 1.5).

- Micropipettes are accurate only between the range of volumes designated by the manufacturer (e.g. $100–1000\,\mu l$ or $0.1–1.0\,ml$). This range is usually on the side and top of the pipette, so check to see if the pipette covers the range required. If it does not, then change the pipette.
- Adjust the volume on the pipette to the volume required (remember that $1.0\,ml$ is $1000\,\mu l$ – see Table 1.1). Do not adjust the pipette beyond its designated range.
- Using your glove-covered hand, hold a suitable tip by the top (wider) end of the plastic and place it on the receiving end of the micropipette (*Note*: Pipettes with different volume ranges require different tips. Please check if you are unsure). Alternatively, pipette tips may be purchased or prepared with all the tips arranged in a box with the open end facing the operator. If this is the case, place the receiving end of the pipette into the open end of a tip, apply pressure to engage the tip with the pipette, then withdraw and use the tip.
- Before drawing up the liquid, hold the pipette in your hand and depress the plunger using your thumb. Notice that there are two stopping points to which the plunger can be depressed; drawing up from the first stop on the pipette will allow the required volume of liquid to be taken up into the pipette tip and depressing to the second stop allows all traces of liquid to be expelled from the pipette tip.
- With the pipette tip on the end of the micropipette, depress the plunger to the first stop; place the tip into the liquid, slowly allowing the plunger to return to the starting position. Check to see that no air bubbles have been drawn into the tip. If an air bubble is present in the tip, discard the liquid back into the stock liquid and start again. Solutions containing proteins and/or detergents are prone to frothing and care must be taken to avoid air bubbles. Despite the care you have taken, sometimes an air bubble will persist, and the most appropriate action is to discard the pipette tip and start again.
- Pipetting solvents can be tiresome because, after taking the liquid into the pipette tip, the solvent may leak from the tip. A simple procedure to avoid this is to take the solvent into the tip and discard the liquid to waste. The next operation of the pipette will take the solvent into the pipette tip and no liquid will drip from the tip.
- Move the pipette to the dispensing container and depress the plunger initially to the first stop, this is followed by moving the plunger to the second stop.

- The pipette tip can then be removed (most micropipettes have a mechanical device to remove the tip) and disposed of according to health and safety regulations.
- Your ability to pipette accurately can be practiced using an empty beaker on a balance and some distilled water. Set the pipette to a volume and dispense water into the beaker on the balance. Note the weight at each addition and after ten additions work out the average weight and the standard deviation (see Chapter 4). The average weight should correspond to the setting on the pipette which will indicate the accuracy of the pipette and the standard deviation will show you how adept you are at dispensing the same volume. If pipettes appear to be inaccurate, they can often be adjusted to the correct calibration by following the manufacturer's guidelines. Repeat the weighing of water until the pipette dispenses the correct volume.

KEY POINTS TO REMEMBER 1.2

- A volume of 0.1 ml of water weighs 0.1 g (100 mg).
- A volume of 1.0 ml of water weighs 1.0 g.
- A volume of 1000 ml weighs 1.0 kg.

1.5.3   Diluting Down from a Stock Solution

A convenient formula can be used to calculate dilutions from a stock solution, but it is important to remember that **all** the unit levels must be the same on both sides of the equation. That is to say, all the units both volume (e.g. ml, μl, nl) and concentration (e.g. M, mM, μM or nM) must be at the same level (see Worked Examples 1.5 and 1.6).

$$C_1V_1 = C_2V_2$$

Initial concentration (C_1) × initial volume (V_1) = final concentration (C_2) × final volume (V_2)

Worked Example 1.5

- You are supplied with 1.0 M stock solution of pyruvate.
What volume of the 1.0 M pyruvate stock is required if you need to prepare 5 ml of a 25 mM solution?

$$C_1V_1 = C_2V_2$$
$$1000 \times V_1 = 5 \times 25 \, (\text{ml})$$
$$V_1 = 5 \times 25 \div 1000 \, (\text{ml})$$
$$V_1 = 0.125 \, \text{ml} \, (\text{or } 125 \, \mu\text{l})$$

Answer: Take 0.125 ml of the 1.0 M stock solution and make this up to 5.0 ml to arrive at a 25 mM solution.

- You are supplied with a 25 mM stock solution of L-lysine.
What volume of the 25 mM L-lysine stock is required if you need to prepare 0.75 ml (750 μl) of a 500 μM solution?

$$C_1V_1 = C_2V_2$$

Convert all the units to the 1 x 10⁻⁶ level (micro) (see Table 1.1)

$$25000 \times V_1 = 500 \times 750 \, (\text{note } C_1 \text{ is 25 mM or 25 000 } \mu\text{M} - \text{see Table 1.1})$$
$$V_1 = 500 \times 750/25000$$
$$V_1 = 375000/25000 = 15 \mu\text{l}$$

The answer is 15 μl or 0.015 ml which should be taken from the stock and made up to 0.75 ml

1.6 Water, Acids and Bases

Most living matter is composed of approximately 70% water and many of its properties, including the formation of H-bonds and the ability to ionize, are fundamental to the correct structure and properties of many biological molecules. Understanding the properties of water and how it interacts with other molecules is fundamental to understanding many topics within biosciences.

The structure of water means that there are electron-rich areas (enriched negative charge) within the molecule and electron-depleted areas (enriched positive charge) called a charge dipole. This dipole of charge is permanent and contributes many of the properties of water because the positive dipole on one molecule of water can interact with the negative dipole on other water molecules (see Figure 1.6a), thus establishing a network of interactions described as hydrogen bonds (H-bonds). In water, most of the molecules are engaged in hydrogen bonding with their nearest neighbours and each bond lasts between 1 and 20 picoseconds (1 picosecond = 10^{-12} s) before the bond is reformed or a new bond is formed with a neighbouring water molecule. Hydrogen bonds are relatively weak bonds (the *bond dissociation energy* is low 23 kJ mol^{-1}) if this is compared to some covalent bonds (470 kJ mol^{-1} to break the O–H covalent bond or 348 kJ mol^{-1} to break a C=C covalent bond). This is also reflected in the greater bond distance of the hydrogen bond compared to a covalent bond (see Figure 1.6a). This means that, as the temperature rises (hence an increase in available thermal energy), the amount of H-bonding decreases. In ice, H_2O molecules form four hydrogen bonds with other H_2O molecules to form a low temperature crystal lattice. The elevated temperature of liquid water results in an increase in the random movement (entropy) of the H_2O molecules and each molecule hydrogen bonds with only 3.4 water molecules.

KEY POINTS TO REMEMBER 1.3

- Hydrogen bonds are important weak bonds that contribute to holding the structures of proteins and nucleic acids together. In most cases, elevated temperatures (e.g. >37 °C) will weaken the H-bonds in the macromolecular structures, causing the structure to unravel and become denatured.
- Hydrogen bonding can be formed between molecules other than water in biological systems. The dipole on electron-rich atoms such as oxygen and nitrogen (relatively negative and thus a hydrogen bond acceptor) and the dipole on the hydrogen attached to oxygen or nitrogen (relatively positive and thus a hydrogen bond donor) may interact if they are within the correct bond forming distance.

1.6.1 Water as a Solvent

Water is a polar solvent because the dipole on water molecules enables water to interact with itself and other charged (*polar*) molecules, allowing them to dissolve (*hydrophilic*) (see Figure 1.6b) compounds. In contrast, nonpolar (*hydrophobic*) molecules cannot interact with water, in the presence of water nonpolar molecules tend to group together. The water molecules surrounding the nonpolar molecules form caged (*clathrate*) structures (see Figure 1.6c).

1.6.2 The Ionization of Water

For every billion water molecules (10^9) there are approximately two molecules that ionize at room temperature. This is a low degree of ionization, but it still represents another important property of water.

Water ionizes into a positively charged hydrogen ion [H$^+$]* and a negatively charged hydroxyl ion [OH$^-$]**

$$H_2O \Leftrightarrow [H^+] + [OH^-]$$

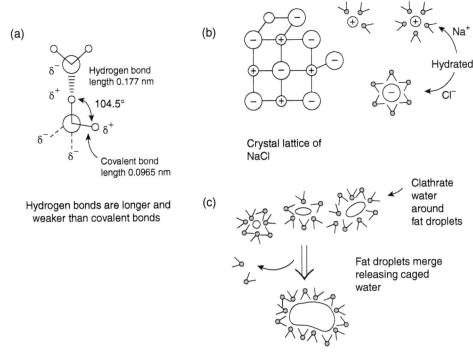

Figure 1.6 *The structure of water (a) contributes to its ability to solubilize ionic compounds (b) but not nonionic compounds (c).*

All reactions can reach equilibrium where the rate of the dissociation is equalled by the rate of association. The equilibrium constant for water K_{eq} at 25 °C is given by the equation.

$$K_{eq} \Leftrightarrow \frac{\left[H^+\right]\left[OH^-\right]}{\left[H_2O\right]} \tag{1.1}$$

In pure water, the molar concentration of water is 55.5 M (divide the weight of 1.01 of water [i.e. 1000 g] by the relative molecular mass of water; Mr = 18), and the K_{eq} for water has been determined experimentally to be 1.8×10^{-16}.

Substitute these values into Eq. (1.1).

$$1.8 \times 10^{-16} \Leftrightarrow \frac{\left[H^+\right]\left[OH^-\right]}{\left[55.5\right]}$$

The ion product of water (K_w) at 25 °C is

$$K_w = \left[1.8 \times 10^{-16}\right]\left[55.5\right] = \left[H^+\right]\left[OH^-\right]$$

$$K_w = 1 \times 10^{-14} \text{ M}$$

The ion product of water (K_w) always equals 1×10^{-14} M. When the concentration of [H⁺] is equal to the concentration of [OH⁻], the solution is at the neutral pH value (7.0).[2]

This value can be calculated when the concentration of [H⁺] is equal to the concentration of [OH⁻].

$$K_w = \left[H^+\right]\left[OH^-\right] = \left[H^+\right]^2$$

Table 1.2 *The pH scale.*

[H⁺] (M)	pH	[OH⁻] (M)
10^0 (1.0)	0.0	10^{-14} (0.00000000000001)
10^{-1} (0.1)	1.0	10^{-13} (0.0000000000001)
10^{-2} (0.01)	2.0	10^{-12} (0.000000000001)
10^{-3} (0.001)	3.0	10^{-11} (0.00000000001)
10^{-4} (0.0001)	4.0	10^{-10} (0.0000000001)
10^{-5} (0.00001)	5.0	10^{-9} (0.000000001)
10^{-6} (0.000001)	6.0	10^{-8} (0.00000001)
10^{-7} (0.0000001)	7.0	10^{-7} (0.0000001)
10^{-8} (0.00000001)	8.0	10^{-6} (0.000001)
10^{-9} (0.000000001)	9.0	10^{-5} (0.00001)
10^{-10} (0.0000000001)	10.0	10^{-4} (0.0001)
10^{-11} (0.00000000001)	11.0	10^{-3} (0.001)
10^{-12} (0.000000000001)	12.0	10^{-2} (0.01)
10^{-13} (0.0000000000001)	13.0	10^{-1} (0.1)
10^{-14} (0.00000000000001)	14.0	10^0 (1.0)

Solving this for [H⁺]

$$\left[H^+\right] = \sqrt{K_w} = \sqrt{1 \times 10^{-14}}$$
$$= 1 \times 10^{-7}$$

The ion product of water is constant which means that when the concentration of [H⁺] is high, there is a corresponding decrease in the concentration of [OH⁻].

The ion product of water (K_w) is the basis of the pH scale, which denotes the molar concentration of [H⁺] in solution.

$$pH = -\log_{10}\left[H^+\right]$$

By taking the negative logarithm, the exponential values (1×10^{-x}) seen above are converted into linear arithmetic numbers (e.g. at neutral pH the [H⁺] = 1×10^{-7} M which is converted to pH 7.0 (see Table 1.2).

- The pH scale is the negative logarithm to base 10 of the hydrogen ion concentration [H⁺]. This means that the values for the molar concentration of [H⁺] decrease by a factor of 10 with each pH number not by a factor of 1.

1.6.3 Acids and Bases

The pH scale outlined in the previous section is a measure of the acidity of a solution describing the molar concentration of [H⁺] in solution. The compounds that can release H⁺ ions into solution are known as acids. In 1923, two scientists Johannes Nicolaus Brønsted (Denmark) and Thomas Martin Lowry (England) independently proposed essentially the same theory of acids and bases. This is now known as the Brønsted–Lowry theory of acids and bases, which proposes that an acid is a proton (H⁺) donor and a base is a proton (H⁺) acceptor. In addition, every acid has a conjugate base and every base a conjugate acid.

$$\text{e.g.} \quad \underset{\text{acid}}{HCl} + \underset{\text{base}}{H_2O} \quad \leftrightarrow \quad \underset{\text{acid}}{H_3O^+} + \underset{\text{base}}{Cl^-}$$

At the same time, another scientist Gilbert Newton Lewis (USA) proposed that acids can be considered as electron pair donors (Lewis acid) and bases as electron pair acceptors (Lewis base).

1.6.3.1 Strong and Weak Acids

Acids are often referred to as being 'strong' or 'weak'. This does not refer to their ability to dissolve other compounds but informs us of their ability to ionize and dissociate in solution.

A strong acid (e.g. hydrochloric acid (HCl), nitric acid (HNO_3) or sulphuric acid (H_2SO_4) will ionize and dissociate at *all pH values* into their conjugate parts of acid and base.

$$\text{i. } \underset{\text{acid}}{HCl} + \underset{\text{base}}{H_2O} \leftrightarrow \underset{\text{acid}}{H_3O^+} + \underset{\text{base}}{Cl^-}$$

$$\text{ii. } \underset{\text{acid}}{HNO_3} + \underset{\text{base}}{H_2O} \leftrightarrow \underset{\text{acid}}{H_3O^+} + \underset{\text{base}}{NO_3^-}$$

$$\text{iii. } \underset{\text{acid}}{H_2SO_4} + \underset{\text{base}}{2H_2O} \leftrightarrow \underset{\text{acid}}{2H_3O^+} + \underset{\text{base}}{SO_4^{2-}}$$

In contrast, the degree of ionization and dissociation of 'weak acids' is dependent upon the pH of the solution in which they are dissolved. Acetic acid (CH_3COOH) is a familiar weak acid which ionizes to varying degrees across the pH scale (see Figure 1.7).

$$CH_3COOH + H_2O \leftrightarrow H_3O^+ + CH_3COO^-$$

Below pH 3.0, when the [H⁺] concentration is high (see Table 1.2), any loss of the [H⁺] on the carboxyl group of acetic acid (–COOH) is quickly replaced. However, above pH 3.0 (i.e. when the [H⁺] concentration in solution decreases rapidly; see Table 1.2), the proton on the carboxyl group of acetic acid can dissociate and go into solution. This leaves a negatively charged carboxyl group on acetic acid and a positively charged proton in solution.

1.7 Buffers

Many biological compounds have weak acidic groups present within their structure. As a result, the surface charge on the molecule will vary according to the surrounding pH. Amino acids, proteins, nucleotides and nucleic acids have weak acidic groups as part of their structures, and the degree of ionization of these molecules is pH dependent. As the integrity and functionality of these molecules depends on the pH, maintaining this optimum pH value is *extremely important* in biology. Buffers are used by scientists to maintain the

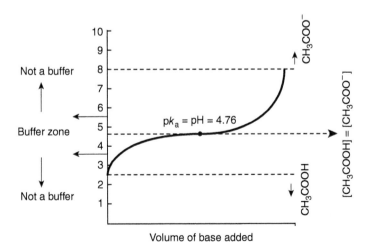

Figure 1.7 *A titration curve for acetate buffer.*

correct structure of biological molecules during *in vitro* experiments. The ideal pH for an enzyme's maximal activity can be determined by assaying the enzyme at different pH values. The pH maximum may not be the cellular pH of 7.4, for example, transglutaminase 2 is a cellular enzyme with a pH maximum of 8.5. This does not mean that the enzyme will be inactive at physiological pH but that it may be catalyzing reactions at approximately 50–70% of its maximum rate.

1.7.1 Preparing a Buffer and Using a pH Meter

There are many different buffers used in bioscience laboratories to preserve and maintain the pH of a solution (some buffers are listed in Table 1.3). A buffer comprises of a weak acid and one of its salts (conjugate base) or a weak base and one of its salts (conjugate acid). Buffers work best when there is a balance between the charged and uncharged species (pK_a) (see Figure 1.7 and Section 1.8). At the pK_a of a buffer, it is able to absorb changes in the concentration of [H$^+$] protons or hydroxyl groups [OH$^-$] without significantly altering the overall pH of the solution. There is a limit to this buffering capacity, which is approximately 1.0 pH unit either side of the buffer's pK_a. If the desired pH value is at the extremes of the buffer's pK_a value, the molarity of the buffer should be increased, or a more appropriate buffer used (see Table 1.3).

- When the desired molarity of the buffer has been decided, the salt(s) which make up the buffer can be weighed out and dissolved (see Section 1.4) in a volume of water (highest purity available) less than the final volume required. Any additions to the buffer that may contribute to the final pH value (e.g. Ethylene diamine tetra acetic acid (EDTA) which is acidic) should be dissolved at this stage before the pH is finalized. When the buffer has dissolved, the pH can be adjusted using a pH meter (see Figure 1.8).
- The probe of the pH meter should be cleaned by rinsing with distilled water and the machine's response calibrated with standard buffer solutions according to the manufacturer's instructions.
- The pK_a value for some buffers can vary with changes in the temperature (e.g. pK_a of Tris at 4 °C is 8.8, but this will change to 8.3 at 20 °C). Most pH meters will also have a temperature adjustment switch. Consider equilibrating the buffer to the temperature (e.g. 37 °C), it will be used at in a planned experiment before adjusting the pH.
- The buffer solution can be placed on a magnetic stirring plate and a suitable magnetic stirring bar 'flea' should be rinsed before placing it into the buffer solution. The stirring rate should not be too fast as the 'flea' may jump and damage the pH probe (particularly important when using glass probes).
- The probe should be rinsed with distilled water prior to being placed in the buffer solution.
- Allow a few minutes for the pH of the solution to stabilize on the pH meter's readout before starting to adjust the pH.
- Remember that buffers are made up of predetermined combinations; for example, sodium acetate buffer comprises a basic component (sodium acetate) and an acidic component (acetic acid). If the desired pH value is overshot, avoid adjusting the pH back to the correct pH with a strong acid or base, for example, hydrochloric acid (HCl) or sodium hydroxide (NaOH). This will use up the buffering capacity thus creating

Table 1.3 *A list of common buffers and their effective pH range.*

Buffer	pK_a	Effective buffer range
Maleate	1.97	1.2–2.6
Acetate/CH$_3$COOH	4.76	3.6–5.6
MES	6.10	5.5–6.7
PIPES	6.76	6.1–7.5
NaH$_2$PO$_4$/Na$_2$HPO$_4$	7.20	5.8–8.0
HEPES	7.48	6.8–8.2
Tricine	8.05	7.4–8.8
Tris/HCl	8.06	7.5–9.0
Na$_2$CO$_3$/NaHCO$_3$	10.33	9.0–10.7

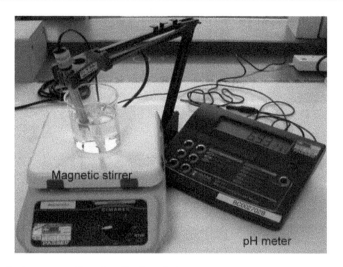

Figure 1.8 *A typical set up to adjust the pH of a solution.*

a less effective buffer. Instead, adjust to the correct pH using a concentrated solution of either one of the buffer's constituents; this will slightly alter the molarity of the buffer but retain the buffering capacity.

- When the correct pH has been attained the pH probe can be removed and rinsed in distilled water before being returned to the pH probe storage solution.
- The buffer can then be made up to volume and stored as required. If the buffer is to be used for chromatography (see Chapter 7), it is advisable to filter the buffer through a 0.2 μm filter to remove any particulate matter and bacteria.
- Overnight storage of the pH probe in a 0.1 M HCl solution will clean and restore most dirty glass pH probes.

1.8   The Equilibrium/Dissociation Constant (K_a) for an Acid or Base and the Henderson–Hasselbalch Equation

A measure of the strength of an acid is the acid-dissociation equilibrium/dissociation constant (K_a) for that acid.

$$\underset{\text{acid}}{HA} + \underset{\text{base}}{H_2O} \leftrightarrow \underset{\text{acid}}{H_3O^+} + \underset{\text{base}}{A^-}$$

The dissociation constant $K_a = \dfrac{\left[H_3O^+\right]\left[A^-\right]}{\left[HA\right]}$

- A strong acid readily dissociates at all pH values and has a high K_a value (HCl: $K_a = 1 \times 10^3$).
- A weak acid has a small K_a value (see Figure 1.7) and its dissociation is pH dependent (CH_3COOH $K_a = 1.8 \times 10^{-5}$).
- To convert K_a values into useable numbers, the pK_a of an acid or base is used which is the negative logarithm of the K_a. In this case, a strong acid will have a low pK_a.

$$pK_a = \log \frac{1}{K_a}$$

18

The Henderson–Hasselbalch Equation.

The quantitative aspects of buffers can be ascertained by the Henderson–Hasselbalch equation, which mathematically links pH with pK_a.

$$pH = pK_a + \log 10 \frac{\left[A^-\right]}{\left[HA\right]} \quad or \quad pH = pK_a + \log 10 \frac{\left[proton\ acceptor\right]}{\left[proton\ donor\right]}$$

- Where $[A^-]$ is the concentration of base and $[HA]$ is the concentration of acid. This is usually a molar concentration but because $[A^-]/[HA]$ is a ratio, other concentration units will also be acceptable. It is this ratio that determines the pH of a solution.
- At the midpoint of the titration curve (see Figure 1.7) when the concentration of $[HA]$ is equal to the concentration of $[A^-]$ the $pH = pK_a$ and this is when buffers are most effective.

$$pH = pK_a + \log 1 = pK_a + 0 = pK_a$$
$$or\ pH = pK_a$$

When $[HA] > [A^-]$ the pH of the buffer is less than the pK_a and when the $[A^-] > [HA]$ the pH is greater than pK_a.
- This $[A^-]/[HA]$ ratio can only be varied within certain limits usually to 1.0 pH unit either side of the pK_a value. This means that there is little or no buffering at the extremes of the buffering range, and it may be necessary to increase the concentration of the buffer to maintain good buffering capacity or switch to a different buffer with a different pK_a.

The Henderson–Hasselbalch equation allows

- Calculation of the pK_a when the pH and molar concentrations of proton donor and acceptor are known.
- Calculation of the pH if the buffer pK_a and molar concentrations of proton donor and acceptor are known.
- Calculation of the molar concentrations of proton donor and acceptor if the pK_a and pH are known.

Worked Example 1.6 The Use of the Henderson–Hasselbalch Equation

Calculate the pH of a solution containing 0.15 M acetic acid and 0.25 M sodium acetate. (For the constituents of a sodium acetate buffer see Table 1.3.)

The pK_a of acetic acid is 4.76

$$\underset{acid}{HA} + \underset{base}{H_2O} \leftrightarrow \underset{acid}{H_3O^+} + \underset{base}{A^-}$$

Acetic acid   acetate

$$\underset{acid}{CH_3COOH} + \underset{base}{H_2O} \leftrightarrow \underset{acid}{H_3O^+} + \underset{base}{CH_3COO^-}$$

$$pH = pK_a + \log_{10} \frac{\left[A^-\right]}{\left[HA^+\right]} \quad or \quad pH = pK_a + \log_{10} \frac{\left[proton\ acceptor\right]}{\left[proton\ donor\right]}$$

$$pH = pK_a + \log_{10} \frac{\left[acetate\right]}{\left[acetic\ acid\right]}$$

$$pH = 4.76 + \log_{10} \frac{\left[0.25\right]}{\left[0.15\right]}$$

$$pH = 4.76 + 0.22$$

$$pH = 4.96$$

1.9 Summary

The foundation of successful experimental work is the correct preparation of solutions. If undertaking a course in the bioscience area, the following skills must be developed to a satisfactory standard, particularly if you wish to pursue a career in laboratory-based bioscience.

- Preparation of molar and percentage solutions.
- Preparation of dilutions from a stock solution.
- The accurate use of pipettes.
- A good understanding of pH and the preparation and use of buffers.

Notes

1 Water from a tap may contain impurities (such as metal ions) that may interfere with an experiment. Water obtained by distillation represents the minimum purity of water that ideally should be used to prepare solutions. Water purification machines that pump tap water through filters can be used to produce distilled and deionized water of varying degrees of purity. The highest purity of water will have a conductivity of $18.2\,\Omega\,ms^{-1}$ ($0.055\,\mu S$).

2 The pH of the cell is approximately 7.0. This is also called the physiological pH value. Water is rarely at pH 7.0 because carbon dioxide (CO_2) dissolves in it to produce carbonic acid, resulting in a slightly acid pH for water of approximately pH 5.5.

$$CO_2 + H_2O \Leftrightarrow H_2CO_3 + H_2O \Leftrightarrow \underset{acid}{H_3O^+} + \underset{base}{HCO_3^-}$$

* [H^+] is effectively a proton, and this is never available in solution. The proton joins up with another water molecule to generate a positively charged hydronium ion [H_3O^+].

** [OH^-] the square brackets indicate that a defined concentration is involved, and this is usually molar concentration unless otherwise stated.

2
MICROSCOPY

2.1 Introduction

The major differences between prokaryotes and eukaryotes are the presence of a nuclear envelope, a cytoskeleton network and intracellular organelles in the latter. The greater complexity of eukaryotic cells enables compartmentalization and greater efficiency of biochemical functions. All living cells are enclosed by a plasma membrane. Most bacteria (prokaryotes) have a peptidoglycan cell wall for structural stability and a membrane that encloses the cytoplasm, the exception being the mycoplasmas, which lack a cell wall altogether. Plant and yeast (eukaryotes) cell walls are made up of other structural components such as cellulose and chitin, respectively. Animal cells have no cell wall whatsoever (Figure 2.1).

Organelles such as nuclei, mitochondria, lysosomes, peroxisomes, rough (RER) and smooth (SER) endoplasmic reticulum and Golgi apparatus are present in all eukaryotes, but not in prokaryotes, whereas plants have additional intracellular structures (e.g. chloroplasts) not present in prokaryotes, yeasts or animal cells. These are summarized in Figure 2.1 and in the bullet point lists below:

The principal organelles/compartments in animal cells and their main functions are:
- Nucleus: DNA replication and transcription.
- Endoplasmic reticulum: Synthesis of membrane proteins and membrane lipids, Ca^{2+} storage and detoxification of xenobiotics.
- Lysosomes: Degradation of macromolecules by acid hydrolases.
- Peroxisomes: Detoxification of hydrogen peroxide.
- Mitochondria: ATP production and Ca^{2+} storage. Semi-autonomous organelles that synthesize some of their own proteins. Mitochondrial DNA and ribosomes are located in the mitochondrial matrix. Double membrane structure with inner membrane folded to form cristae.
- Free polyribosomes: Synthesis of cytoplasmic, nuclear, mitochondrial and peroxisomal proteins.
- Glycogen granules: Storage form of glucose (*replaced by starch in plants).
- Cytoskeleton: Interlinked network of three sets of protein filaments known as microtubules, microfilaments and intermediate filaments, which are collectively involved in the regulation of cell division, cell movement, cell shape and intracellular transport. For example, in mitosis, microtubules from the mitotic spindle (which aligns and separates chromosomes during metaphase and anaphase, respectively) and microfilaments form the contractile ring (which generates the cleavage furrow leading to the formation of two identical daughter cells).
- Centrioles: a pair of these forms the poles of the mitotic spindle, which act as nucleation sites for microtubule assembly both during cell division (spindle microtubules) and the subsequent interphase stage (astral microtubules).
- Cytosol: A mixture of soluble proteins that carry out a variety of structural (e.g. cytoskeletal networks) and enzymatic roles within the cell (e.g. enzymes involved in glycolysis).

In addition to the above structures, additional features are found in plant cells, including:
- A large central vacuole: Uptake of water, storage of materials and degradation of macromolecules by acid hydrolases.
- Chloroplasts: Photosynthesis. Semi-autonomous organelles that synthesize some of their own proteins. Triple membrane (inner, outer and thylakoid) structure.
- Cellulose cell wall: Structural rigidity.
- Starch grains: A storage form of glucose.

Basic Bioscience Laboratory Techniques: A Pocket Guide, Second Edition. Philip L.R. Bonner and Alan J. Hargreaves.
© 2022 John Wiley & Sons Ltd. Published 2022 by John Wiley & Sons Ltd.

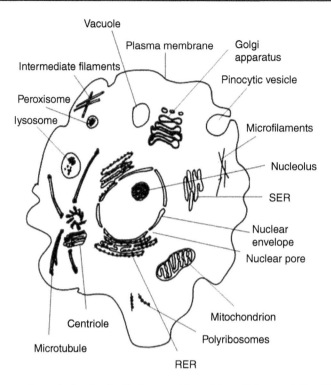

Figure 2.1 *Diagram of a typical animal cell. Source: Adapted from Nelson and Cox (2012). Lehninger: Principles of Biochemistry 6th Ed 9. WH Freeman.*

Ultrastructural detail of all the above macromolecular structures can be seen using transmission electron microscopy. Although structural detail of macromolecules cannot be visualized directly in the light microscope, it is possible to demonstrate the location and distribution of many macromolecules within cells and tissues with the aid of histochemical staining (see later). In combination with biochemical subfractionation techniques (see Chapter 5, Section 5.5), microscopy has played a central role in the study of cell structure and organelle function.

Student exercise

- Draw a labelled diagram of a typical animal cell, indicating the main structural components.
- List any additional structures that are absent from animal cells but present in (a) plant cells and (b) yeasts.
- Indicate which of the structures found in eukaryotic cells are absent from prokaryotes.

2.2 Microscopes – General Principles

All microscopes consist of a coordinated system of lenses providing a magnified image of the specimen. They are invaluable tools in the study of the structure, organization and migration of cultured animal and bacterial cells. Microscopy can also be used to determine tissue organization and the viability, distribution and function of cells within tissues as well as the identification, organization and quantification of organelles and macromolecules within cells and tissues. Microscopy therefore has applications in all of the bioscience-related subjects including microbiology, cell biology, biochemistry, biomedical sciences, forensic science, pathology, toxicology, plant biology, and so on, and is particularly useful at providing supporting data for molecular studies of cellular components.

The type of microscope needed for a particular task depends on the level of detail required. The main differences between the different types of microscopes are the wavelength of electromagnetic radiation (see Chapter 3, Section 3.2) used to obtain the image, the nature and arrangement of the lenses and the methods used to view the image. In this chapter, we will look at the principles of light and electron microscopy, but before that we will consider some general principles applicable to all types of microscopy.

2.3 Principles of Image Formation

There are three basic principles of image formation: resolution, magnification and image formation.

i. *Resolution*: is the minimum distance apart that two points can be distinguished as two separate objects. The resolving power of the unaided eye is approximately 0.2 mm, while that of the light microscope is approximately 200 nm, whereas the transmission electron microscope (TEM) can achieve a resolution of 0.1 nm.
ii. *Magnification:* A common principle in all systems is that the objective lens forms an intermediary image, which becomes the object of the projector lens, which then produces the final image. For example, in a basic laboratory-based light microscope, the projector lens is the eyepiece, which projects an image on to the retina, a camera or a detector that provides an image suitable for viewing (e.g. on a monitor).
iii. *Contrast:* The level of contrast determines the visibility of the specimen caused by relative differences in light (or electron beam) intensity as it emerges from different areas of the specimen. Contrast can be enhanced by using chemical stains or a range of contrast enhancement techniques which will be discussed later. Specimens can contain amplitude objects or phase objects (see Figure 2.2). Amplitude objects decrease intensity by scattering or absorbing light, whereas phase objects induce phase changes in the light passing through them which must be converted to intensity changes in order to see enhanced contrast. Intensity changes are then observable by eye.

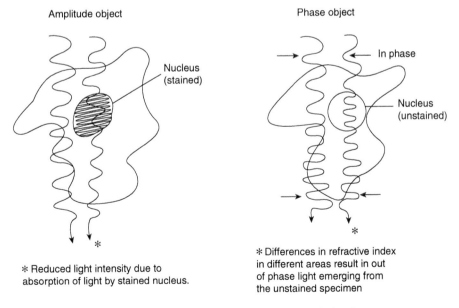

Figure 2.2 Schematic representation of amplitude and phase objects in light microscopy.

2.3.1 Units of Measurement

The units of measurement you are most likely to come across in microscopy and their symbols are the metre (m), the centimetre (cm), the *millimetre* (mm), the *micrometre* (μm), the *nanometre* (nm) and the angstrom (Å), with the most common ones shown in italics (see Table 2.1). The level of detail discernible by different microscope techniques is summarized in Table 2.2.

Student exercises

Memorize the units of measurement commonly used in microscopy and their interrelationships with each other.

Convert the following values:

0.2 mm into nm
Convert 3 cm into μm
Convert 50 Å into nm
Convert 11 mm into μm

Table 2.1 *Conversion chart for units of measurement (distance) in microscopy.*

Unit	Equivalent in other units					
	m	cm	mm	μm	nm	Å
m	—	10^2	10^3	10^6	10^9	10^{10}
cm	10^{-2}	—	10	10^4	10^7	10^8
mm	10^{-3}	10^{-1}	—	10^3	10^6	10^7
μm	10^{-6}	10^{-4}	10^{-3}	—	10^3	10^4
nm	10^{-9}	10^{-7}	10^{-6}	10^{-3}	—	10
Å	10^{-10}	10^{-8}	10^{-7}	10^{-4}	10^{-1}	—

Table 2.2 *Relationship between object size ranges, microscopic technique required for detection and observation.*

Observable detail	Size range
Atoms[a]	0.1–1 nm[a]
Proteins[b]	1–10 nm[b]
Macromolecular structures (e.g. cytoskeleton), ribosomes, small membrane vesicles	10–100 nm[b]
Mycoplasmas and small organelles (e.g. mitochondria)	0.1–1 μm[b]
Bacterial cells and larger organelles (e.g. chloroplasts and nuclei)	1–10 μm[c]
Eukaryotic cells	10–100 μm[c]
Tissue organization	100 μm–1 mm[c]

[a] Observable only by electron microscopy.
[b] Observable by electron microscopy; detectable by light microscopy but only when using histochemical staining techniques.
[c] Observable by all techniques with appropriate chemical stains or optical enhancement methods.

2.4 Light Microscopy

2.4.1 Bright Field Microscopy

The earliest microscopes had a single convex lens and are referred to as simple microscopes. Modern light microscopes are called compound microscopes because they use a combination of biconvex objective and eyepiece lenses to provide a magnified image.

The first compound microscope was developed by Janssen in 1595 with the capability to magnify an object from 10 to 30 times. The first observations at cellular level were made by Robert Hooke around 1665 who coined the term 'cell' for the cell-like units observed in cork tissue. However, the first observations of *living* cells (including bacteria, protists, algae, sperm cells and ciliates) were published by Antonie van Leeuwenhoek between 1673 and 1700 using a simple microscope capable of magnifications up to 200 times.

Since that time, major improvements to the compound microscope have led to superior magnification and image quality compared to the simple microscope. A typical lens arrangement for a compound microscope is shown in Figure 2.3.

2.4.1.1 Specimen Preparation

If a non-living specimen is to be observed, it is typically fixed using a solution of formalin (4% (w/v) *p*-formaldehyde), after which it is dehydrated and embedded in paraffin wax or a hard epoxy resin. Other methods of fixation may be used in specific circumstances. For example, cultured cell monolayers can be fixed at $-20\,°C$ with 90% (v/v) methanol and blood smears or bacteria are sometimes fixed to a slide by gentle heating above a Bunsen flame, after which they can be stained directly.

If tissue sections are being stained, embed the sample in paraffin wax or hard resin, which allows it to be sliced into thin sections of a few microns in thickness, using a microtome. The sections are carefully transferred on to a glass slide before being incubated with chemical stains to enhance contrast on viewing in the light microscope. Some chemical stains that are commonly used in light microscopy are described in Table 2.3.

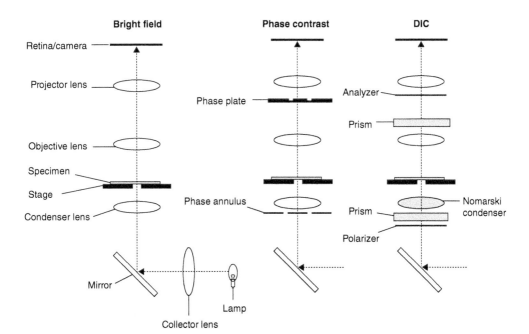

Figure 2.3 *Lens arrangements in bright field, phase contrast and DIC microscopy.*

Table 2.3 *Examples of chemical stains and their uses.*

Name (other names)	Examples	Staining pattern
Basic (cationic) dyes	Hematoxylin Crystal violet Giemsa Safranin	Stain basophilic structures blue or purple (e.g. heterochromatin). Commonly used for nuclear staining. Can aid in the identification of mitotic cells by the characteristic staining patterns due to altered chromosome distribution.
Acidic (anionic) dyes	Eosin Acid fuchsin Fast green	Stain structures that contain positive charge (e.g. proteins, cytoplasm).
Protein binding dyes	Coomassie Brilliant Blue	Used to stain proteins in electrophoresis gels and thus stain the whole cell blue.
Metallic stains	Silver nitrate	Used to stain neurofibrils in neurons giving better contrast than H & E
Lipid stains	Sudan black	Used to stain phospholipids and other lipids. Can be used to identify granules in neutrophils on blood and bone marrow smears.
General stains	Toluidine blue	General cell stain used in plastic-embedded sections. Some formulations are used to stain carbohydrate polymers in cell walls and cartilage.
Immunohistochemical stains	Horseradish peroxidase-conjugated antibodies	Specific antigen localization following incubation with chromogenic substrate that produces coloured insoluble deposits around antigen location.

Examples of Coomassie Brilliant Blue (CBB) and hematoxylin/eosin (H & E) staining of cultured cells are shown in Figure 2.4. Note that CBB stains protein throughout the cell, whereas H & E stains mainly the nucleus blue or purple (dark) and the cytoplasm pink (light). In the case of unfixed or living specimens, samples are visualized directly in the light microscope taking advantage of any natural pigments (e.g. chlorophyll) or using enhanced contrast optics (discussed later).

Many staining procedures involve a combination of two or more dyes, one often acting as a counterstain. Specimen slides are often placed on slide racks or dipped in slide staining jars during the various staining steps. Two commonly used staining protocols are outlined in the next section.

2.4.1.2   Gram Stain for Bacteria

Method:

1. A heat-fixed smear of bacterial cells is flooded with crystal violet for 30 seconds before a quick rinse with distilled and deionized water.
2. It is then flooded with Gram's iodine for 60 seconds then rinsed with distilled and deionized water.
3. Next the colour is removed with 95% (v/v) ethanol and then rinsed again in distilled and deionized water.
4. The specimen is then counterstained with safranin for 60 seconds, before being rinsed with distilled and deionized water, carefully blotted dry using tissue paper prior to examination with the aid of a light microscope using an oil immersion lens.

Interpretation:

Gram-positive bacteria stain blue/violet because they retain the crystal violet after washing with 95% (v/v) ethanol, as they have a thicker cell wall than Gram-negative bacteria. The latter lose the dye during the ethanol wash and are eventually stained pink by the safranin counterstain, which is overwhelmed by

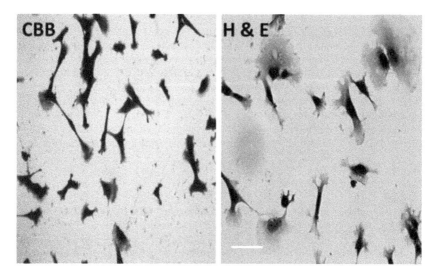

Figure 2.4 *Images of chemically stained cultured cells. Cultured epithelial cells were fixed and then stained with CBB (left panel) or H & E (right panel). Note that CBB shows prominent staining of proteins in both nuclei and cytoplasm, whereas hematoxylin stains mainly nuclei (dark) and eosin counterstains the cytoplasm (light). Bar represents approximately 20 μm.*

the darker crystal violet stain in Gram-positive cells. Apart from revealing the presence of Gram-positive (purple), Gram-negative (pink) or both types of cells, the shape of cells present (rod-like shape for Bacilli and ovoid or spherical shape for cocci, etc.) can also be distinguished in the light microscope. Care needs to be taken not to overstain or decolourize the specimen for too long, as this could lead to false-positive or false-negative results, respectively. Gram staining is almost always the first step performed in identification of bacteria from a variety of isolates taken from hospital patients, industry, food, food preparation areas and the environment.

2.4.1.3   Hematoxylin/Eosin Staining of Paraffin Wax Sections

Method:

1. Thin sections are deparaffinized (e.g. by gentle warming and a 3-minute wash in xylene).
2. They are then rehydrated by consecutive washes in absolute ethanol (3 × 3 minutes), 95% (v/v) ethanol (1 × 3 minutes) and 80% (v/v) ethanol (1 × 3 minutes) finishing with a 5-minute wash in distilled and deionized water.
3. Specimens are then stained for 3 minutes in hematoxylin/eosin, rinsed in distilled and deionized water, then incubated in tap water for 5 minutes to allow colour development.
4. Destaining is achieved by dipping quickly 8–12 times in acid ethanol (2 ml concentrated HCl made up to 100 ml with 95% (v/v) ethanol), followed by two 1-minute rinses in tap water and a 2-minute rinse in distilled and deionized water.
5. Excess water is carefully blotted away from the slide before dipping in the eosin counterstaining solution for 30 seconds, followed by three 5-minute washes in 95% (v/v) ethanol and three-minute washes in absolute ethanol.
6. After three 15-minute washes in xylene, a drop of xylene-based slide mountant (e.g. Permount) is applied on to the coverslip, avoiding bubbles.
7. The coverslip is carefully placed on the slide such that the mountant spreads evenly between it and the glass slide. After drying, the specimen is ready to view in the light microscope.

Interpretation:

Nuclei stain blue/violet with hematoxylin and the cytoplasm stains pink with eosin in most eukaryotic cell types. Mammalian red blood cells can be distinguished in tissue sections by their ellipsoid shape, small size and pink colour (due to the absence of nuclei on maturation) and their presence in a clear area (capillary lumen) surrounded by an ordered cell layer (capillary endothelial cells). It is possible to observe tissue damage and tissue organization using this microscopy technique which allows the identification of different cell types, e.g. by their distinctive shape, size or staining pattern. Hematoxylin and eosin (H & E) staining can also be utilized to estimate mitotic index (i.e. the percentage of cells showing condensed chromatin arrangements corresponding to chromosomes at various stages of cell division) and to identify potential tumour cell growths, which would be expected to have a higher mitotic index than normal tissue.

2.4.2   Phase Contrast Microscopy

This method utilizes contrast enhancement optics so that unstained fixed or living tissue can be observed. Bright field microscopy detects reduced amplitude of light to generate contrast, whereas phase contrast microscopy detects changes in refractive index. Conversion of a microscope from bright field to phase contrast can be achieved by the inclusion of a 'phase ring' and a 'phase plate' in the light path (see Figure 2.3). Diffracted light is spread evenly, whereas a hollow cone of undiffracted light is passed through the specimen. The phase optics allow in-phase and out of phase light to be recombined, resulting in enhanced contrast. A potential drawback with this method is the appearance of a bright ring particularly around small spherical objects, sometimes referred to as the 'halo effect' (see Figure 2.5).

2.4.3   Differential Interference Contrast

Differential interference contrast (DIC) uses special optics to detect changes induced in the orientation of plane polarized light by birefringent objects. DIC utilizes the interference between separated wave fronts, giving highest contrast where the differential is greatest. Conversion from bright field to DIC requires the use of a specialized 'Nomarski' condenser unit, and the addition of an upper prism, a secondary prism,

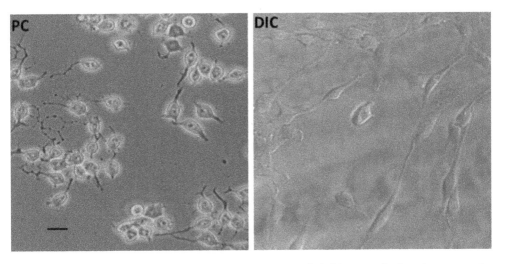

Figure 2.5 *Images of cultured live neuronal cells. Shown are digital images of cultured neurons using phase contrast (PC; left panel) or differential interference contrast (DIC; right panel) optics. Note that many cells in the left panel are surrounded by clear areas due to the halo effect, which is reduced in DIC. Bar represents approximately 20 μm.*

an analyzer and a polarizing filter, as indicated in Figure 2.3. The images produced appear to be three-dimensional due to a shadow-cast effect of polarized light. DIC has several advantages over phase contrast, including the abolition of the halo effect (see Figure 2.5) and objective lenses with a shallower depth of field, enabling optical sectioning.

2.4.4 Fluorescence Microscopy

This specialized technique allows the detection of small molecules using fluorescent probes (e.g. antibodies labelled with fluorophores, DNA and intracellular calcium ion binding dyes, etc.). This facilitates a range of cellular assays, some of which are shown in Table 2.4 and Figure 2.6. The principles of immunofluorescence

Table 2.4 *Fluorescence-based cellular assays.*

Fluorophore or fluorophore-labelled probe	Application
Calcium binding dyes (e.g. Fura-2 and Quin-2)	Measurement of intracellular Ca^{2+} concentration
DNA-binding dyes (e.g. DAPI and propidium iodide)	Nuclear staining, identification of cells at different stages of the cell cycle due to characteristic changes in chromatin organization. Cell proliferation assays. Cell death assays.
Carboxymethyl fluorescein succinimidyl ester (Note that fluorophores other than fluorescein, such as rhodamine and AlexaFluor dyes, can also be used in this way and sometimes with other attachment groups)	Stains the cytoplasm in viable (healthy) cells. Can be used in combination with DNA-binding dyes such as propidium iodide to identify living and dead cells by live cell imaging techniques. Can be used as a cell tracking dye (e.g. in cell migration assays). Can be used to covalently label antibodies and other proteins for use in immunofluorescence staining and a range of other assays, such as detection of protein targets in cultured cells and tissue slices.
Dihydroxy fluorescein diacetate	Measurement of reactive oxygen species in assays of oxidative stress.
D_2R (Rhodamine 110, bis-(L-aspartic acid amide), trifluoroacetic acid salt	Cell permeable substrate used to measure caspases 3 and 7 activity in living cells. Caspase-mediated cleavage releases rhodamine, which produces intracellular fluorescence indicative of the level of activity *in situ*.
N-succinyl-L-leucyl-L-leucyl-L-valyl-L-tyrosine-7-amido-4-methylcoumarin	Cell permeable substrate for the measurement of calpain activity in living cells. Calpain-mediated cleavage releases amidomethyl coumarin, which produces intracellular fluorescence indicative of the level of activity *in situ*.
JC-1	Accumulates in mitochondria, where it can be used to monitor membrane potential.
Fluorescein-labelled cadaverine	Cadaverine is a monoamine, the covalent incorporation of which into cellular proteins is mediated by the enzyme transglutaminase. This assay gives a measure of transglutaminase activity *in situ*. A similar approach can potentially be applied to the measurement of other enzyme activities for which fluorescent substrates are available.

Shown are selected examples of fluorescent molecules and some applications in which they can be used.

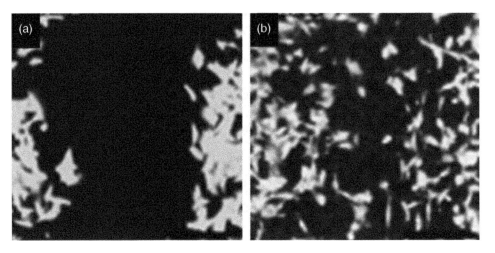

Figure 2.6 *Images of cells stained with carboxymethyl fluorescein succinimidyl ester (CFSE). In panel A is shown a monolayer of CFSE-labelled epithelial cells following the application of a scratch to remove cells from the central zone (dark). Panel B shows the same cells after 8 hour incubation, during which they have begun to migrate from the edges into the gap.*

microscopy are discussed in Chapters 3 and 9. Fluorescence is a process where molecules with the correct chemical structure absorb energetic light at a specific wavelength number called the excitation wavelength (photons with short wavelength number are more energetic than photons with a high wavelength number nm) and the molecule will re-emit (emission wavelength number nm) the light at a longer wavelength number (see Chapter 3, Section 3.3.1). Fluorophores are molecules that can be excited to a higher energy state by irradiation with an appropriate wavelength of light (excitation wavelength), subsequently releasing some of the absorbed energy in the form of fluorescence at a longer wavelength (the emission wavelength). Fluorophores can be covalently attached to a multitude of probes to detect molecules that bind the probe. For example, staining of tissue sections with fluorescently labelled antibodies enables the localization of antigens by immunofluorescence staining (see Chapter 9 for details). In contrast to light microscopy, fluorescence corresponding only to the target molecule of interest is detected against a dark background. For example, fluorescently labelled antibodies to tubulin would reveal a fluorescence pattern that aligned with the microtubule network, whereas propidium iodide staining would give a fluorescence pattern corresponding to DNA in the nucleus.

Visualization of fluorescence (see Chapter 3) requires the use of a fluorescence microscope, which normally uses a mercury lamp and a set of wavelength filters and beam splitters to irradiate the specimen evenly through the objective lens (acting as the condenser lens) (see Figure 2.7). Most of the light passes through the specimen but some is absorbed, releasing energy in the form of fluorescence at a longer wavelength number, which passes back through the objective lens and straight through the beam splitter up to the eyepiece.

A major problem with this approach can be interference from out-of-focus fluorescence, which gives a fuzzy background. This can be eliminated using a confocal laser scanning fluorescence microscope, which scans the specimen at different heights through a narrow plan of focus using precise laser beams to eliminate stray light interference.

2.4.5 How to Use a Light Microscope

Rather than getting a big enough magnification, the main challenges in light microscopy are obtaining sufficient contrast and resolution, locating the focal plane and recognizing the structures being viewed. The microscope you are most likely to come across in your studies is the bright field microscope.

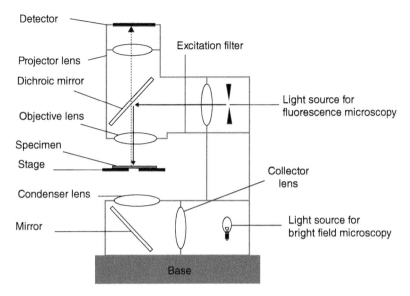

Figure 2.7 *Lens arrangement and light path in epifluorescence microscopy.*

2.4.5.1 General Features of Microscopes

In a conventional laboratory bright field microscope, the objective and eyepiece lenses are located above the specimen stage and the light source below it, as indicated in Figure 2.8. In the case of the inverted microscope, typically used in cell culture work, the objective lenses and light source are below the sample stage, but the eyepiece lenses remain above it. A good microscope has a built-in illuminator, an adjustable condenser with an aperture diaphragm (which can be reduced in diameter to increase contrast up to a point), a mechanical stage and a binocular arrangement of eyepiece lenses.

The purpose of the condenser is to focus an even beam of light on the specimen, which, after passing through the objective and eyepiece lenses, is significantly magnified by the time it reaches the eye. If you are undertaking bright field microscopy and there are adjustable settings on the condenser, make sure that it is set to bright field. Some condensers are fixed whereas others can be focused to adjust the quality of the light. If focusable, the best position for the condenser is as close as possible to the microscope stage. The condenser is usually fitted with an aperture diaphragm, which can be opened to increase brightness at the expense of contrast and vice versa. When closing the aperture to increase contrast, a point is reached where the image becomes distorted. If the condenser requires adjustment, you should follow instructions given in the class or in the instrument manual, as precise details may vary from one microscope to another.

2.4.5.2 Focusing a Specimen

- Place the specimen slide carefully on the microscope stage with the light source switched on. There may be some slide clips or a groove to hold the slide in position.
- Make sure that the coverslip that covers the specimen is facing towards the objective lens, as high-power lenses are unable to focus through the relatively thick glass slide. Place the sample in the light beam and bring it into focus using a low magnification objective lens (typically ×10 or ×4).
- This is relatively easy to do if the sample is sectioned and stained but if using microbial cells, it can be useful to look for the edges of smears, where cell density is usually higher.

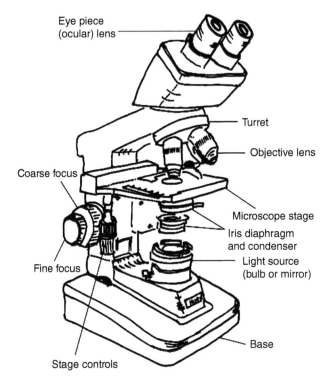

Figure 2.8 *Diagram of a typical light microscope.*

- In a good quality microscope, the stage can be moved mechanically, enabling precise movements of the slide without touching it.
- Start with the specimen out of focus so that the stage and objective need to be moved closer together. The first surface coming into focus will be the top of the cover slip, although this is not always applied over smeared samples, then the lower surface of the cover slip which is the plane of focus for the sample.
- Once the sample is in focus at low magnification, it is now possible to swing the higher magnification lens into position and refocus to obtain the higher magnification image. The sample may need to be refocused and recentred although with good quality lens systems, the specimen should come almost into focus as the lenses are changed. It will also be necessary to adjust the light as higher magnification lenses to cover a smaller field of view and, therefore, less light passes into the lens at a given light intensity. Thus, with a low power lens, the illumination intensity can be decreased and vice versa for high power objectives.
- The contrast and light intensity can then be adjusted, and the sample position moved to obtain an optimal image. If oil immersion is required (e.g. with high power lenses ×100), a drop of oil is placed on top of the cover-slip area above the sample before bringing it into contact with the oil immersion objective lens while focusing. The oil has the same refractive index as the glass slide, which enhances resolution. The oil immersion lens is essential for microbiological samples, especially bacteria.
- Remember that there are usually crude (outer) and fine (inner) focus controls, the latter being used at higher magnifications. However, the high-power lenses need to be much closer to the specimen; care is therefore needed to prevent contact with the coverslip, which could jam the slide in position.
- Remember also that the coverslip should be fixed in position, particularly if the sample is covered by a layer of liquid (e.g. in fluorescence microscopy).

HOW TO LOOK AFTER A MICROSCOPE:

All of the parts of a good quality microscope are expensive, so care is needed when handling and using a microscope. In particular:

- When moving a microscope, always hold it firmly by the stand.
- When unplugging the light source, always hold the plug (not the cable).
- As bulbs are expensive and have a limited life span, the illuminator should be turned off as soon as you have finished.
- Never use immersion oil on a dry lens, as the image will be distorted.
- Always clean the lenses (gently with proper lens tissue only!) and the stage before leaving the microscope.
- An appropriate lens cleaning fluid may be applied to remove dried material, but some organic solvents may damage lens' surfaces (see the manufacturer's recommendations).
- Place the designated dust cover on the microscope when it is not in use.
- Try to focus gradually and not too quickly, without forcing the focusing controls. If you encounter resistance, you have probably gone too far past the specimen and need to start again.
- If moving between samples under two different coverslips on the same microscope slide using oil immersion lenses, avoid the risk of scratching the lens by lowering the microscope stage before moving the slide to the new position and refocusing.
- Never remove the lenses from the microscope unless you know what you are doing or are instructed to do so.

2.4.6 Data Analysis and Presentation

2.4.6.1 Drawing and Labelling a Microscope Image

- Remember to give each image a clear descriptive title.
- If drawing the image, draw representative areas of cells rather than every detail. Label the major structures or cell types visible in the field of view.
- If drawn by hand, indicate the magnification at which the specimen was viewed or, if using a camera image, include the final magnification (including image enlargement) or a calibrated scale bar to give a clear idea of dimensions and distances.

Magnification can be determined by multiplying the magnifying powers of the eyepiece (ocular) and the objective lens used to view the specimen (see Worked Example 2.1).

Worked Example 2.1 Calculation of image magnification

The magnification observed using a ×10 eyepiece lens and a ×40 objective lens would be:

$$10 \times 40 = 400 - \text{fold magnification}$$

Note that, if making prints from a negative, the enlargement factor of the print from the negative also needs to be taken into account when calculating the final magnification. Many microscope cameras are digital, and the camera software allows the insertion of a scale bar on to the image, which would adjust according to the final image size.

2.4.6.2 Making Size Measurements Using the Light Microscope

Microscopes may be fitted with a scale (reticule) in one eyepiece lens, which can be turned and aligned with any focused object to obtain an estimate of its size. The size of the object is initially noted as the number of ocular divisions it spans. This number is then multiplied by a conversion factor which is different for each objective lens and obtained by calibrating it against another micrometre scale etched onto on a glass slide and brought into focus on the microscope stage (see Worked Example 2.2).

A typical eyepiece reticule has a scale of 50–100 equal divisions, which can be brought into focus by adjusting the eyepiece focus if necessary. A typical stage micrometre scale is 2 mm long and it has areas with specific divisions (e.g. 0.01 mm or 10 µm). For calibration, these two scales are aligned as shown in Figure 2.9.

Worked Example 2.2 Calibration of lenses with an eyepiece reticule and stage graticule

- Let us suppose that the alignment shown in Figure 2.10 was carried out using a ×100 objective lens.
- If each stage graticule division corresponds to 10 µm then, at this magnification, one eyepiece unit would be equivalent to 1 µm (10 nm × 100).
- At lower magnification, this value would increase as the eyepiece reticule scale stays the same while the objective lens magnification and, therefore, the stage graticule image size decrease.
- Thus, with ×40 and ×20 objective lenses on the same microscope, one eyepiece reticule unit would be 2.5 µm and 5 µm, respectively.

When making single estimations of length or distance, you should bear in mind the fact that the actual limits of resolution of a dry objective lens (up to ×40) and an oil immersion lens (×100) on a basic laboratory instrument is unlikely to be better than 1 µm or 0.5 µm, respectively. Values should therefore be rounded up or down to the nearest whole number. For example, if you calculate the size of an object to be 4.55 µm, you should round up this value to 5 µm (*not* 5.0 µm, which suggests an accuracy of 0.1 µm). Of course, accuracy could be improved by making several measurements of independent objects and calculating the mean × SD (see Chapter 4), in which case the use of precise estimates to one or two decimal places would be more meaningful.

2.4.6.3 Quantification of Specific Features in Microscope Specimens

In a laboratory class, you may only be required to make a few observations or measurements (e.g. number of mitotic cells, number of red blood cells per unit area, etc.) on one or two specimens. However, with a small number of microscope fields, you cannot be sure that you are viewing a representative area of tissue. In fact, to produce publishable research data, multiple fields in several independent specimens would be analyzed in order to take experimental variation into account. Within the time constraints of a laboratory class, comparison of your observations with those of other groups will give you a better idea of the reliability of your observations. Obviously, major differences in measurements between individuals may be due to inconsistencies in the method of measurement especially if the same specimens were analyzed. Alternatively, if all of the samples are different sections, differences could also be due to the orientation of the sections cut.

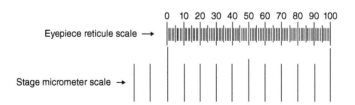

Figure 2.9 *Eyepiece reticule calibration. Source: Adapted from Nelson and Cox. Lehninger: Principles of Biochemistry 5th Ed 9, WH Freeman.*

(a)

(b)

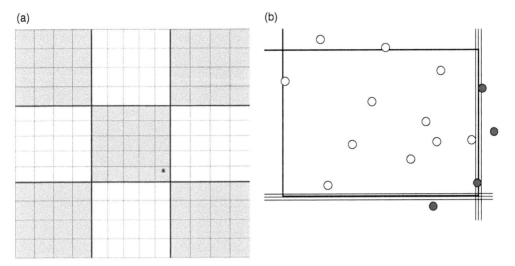

Figure 2.10 *(a and b) Counting cells in a haemocytometer chamber. Shown is a schematic representation of a haemocytometer chamber grid viewed in the light microscope. In (a), each large shaded counting area covers 1 cm², having a volume of 0.1 µl. There are two different sizes of small squares in the counting areas. Those within the central large area have the dimensions 0.2 mm × 0.2 mm and there are 25 per large area. Those in the outer 4 large areas are 0.25 mm × 0.25 mm and there are 16 per large area. The dark grid lines in the diagram usually appear as a triplet of parallel grid lines in the microscope field of view. (b) The enlargement of the single small square marked with an asterisk shows that cells appear as spheres. Only cells within the boundaries of the large squares bordered by these triple lines should be counted. Cells that are outside or that overlap with the middle line of the triple line boundaries (shaded) should be excluded from the counts. Source: Adapted from Nelson and Cox Lehninger: Principles of Biochemistry 6th Ed 9, WH Freeman.*

2.4.6.4 Application of Microscopy in Cell Counting

A simple example of quantification using the light microscope is the use of a hemocytometer chamber to count cells in a cell suspension. In this instance, the cells are suspended in an isotonic medium and a small volume of the suspension applied under the coverslip of a haemocytometer chamber. On focusing the microscope, an image like the one in Figure 2.10 will appear. In the actual focused slide, triple lines (depicted by thicker lines in the diagram) delineate the major areas from which cells are to be counted.

Cultured cells are normally suspended in a medium, diluted in a buffer or a medium, before 10 µl are applied to the chamber beneath the coverslip. If the cell number is overwhelming, the suspension can be diluted (1/10 to 1/200) and recounted. Cell density (i.e. the number of cells per ml) can be estimated in one of the following ways (see Worked Example 2.3).

a. For most mammalian cells in suspension (e.g. cultured cells), count all of the cells lying within five large areas, then estimate the cell density (cells per ml suspension) by applying the formula:

Mean count per large area × dilution factor (typically 10 or 20) × 10^4 (volume factor)

b. For leucocytes, count the cells in the four large corner squares, then calculate the number of cells per ml from the formula:

Average cell counts per large corner square × dilution factor (typically 20) × 10^4

c. For a suspension of red cells, count the erythrocytes in five *small* squares in the middle large area, estimating the cell density (cells per ml) by the formula:

(Total counts from 5 squares × 5) × dilution factor (typically 200) × 10^4

Worked Example 2.3 Estimation of cell density from counts in a haemocytometer chamber

Estimate the cell density of a suspension of white blood cells which, when diluted 10-fold gives counts of 55, 67, 49 and 70 cells in the four outer large counting areas of a haemocytometer grid.

- First of all, work out the average cell count per large counting area, as follows:
  (55 + 67 + 49 + 70)/4 = 60.25 = 60 (rounded down to nearest whole number)
- Incorporating this value into the equation described above for leucocytes:
  Cell density = $60 \times 10 \times 10^4 = 6 \times 10^6$ cells/ml

Student exercises

A suspension of cultured white cells diluted 1/20 gave counts of 28, 36, 42 and 33. Calculate the original cell density. A suspension of cultured human cardiomyocyte cells diluted 1/10 gave counts of 48, 55, 42, 53 and 63. Calculate the original cell density.

2.5 Electron Microscopy

As indicated in the general microscopy section, the electron microscope (EM) offers better resolution than the light microscope and a higher magnification range. The two main types of EM are known as transmission (TEM) and scanning (SEM), which use electron beams differently to obtain images with different characteristics, as indicated in Table 2.5.

- When viewing images of thin sections in the TEM, you should bear in mind that the profiles of larger organelles observed will be affected by the thickness of the section, the location of specific organelles and by the angle of the section relative to the organelle profile.
- Thus, a sausage-shaped mitochondrion might appear spherical if caught in cross section, more sausage shaped in a symmetrical longitudinal section or something in between. Additionally, two spherical profiles appearing side by side in a thin section could be two separate mitochondria or the same long mitochondrion moving in and out of the plane of the section.
- Changes in cell structure can be estimated using morphometric measurements to count the number of profiles corresponding to particular organelles per unit area. Bearing in mind what has been said, it is unlikely that a single section can be representative of the whole cell.
- Therefore, many images should be analyzed from different parts of the specimen and from several specimens under each experimental condition in order to carry out statistically valid estimations of organelle distribution.
- The most common approach to quantify changes in organelle composition or volume relative to total cell volume is to superimpose a grid template with grid points at a fixed distance apart (e.g. 1 cm). The organelle or structure in contact with each point of the grid is recorded and can be estimated as a percentage of the total volume (grid points).
- Clearly this is only a very rough estimate and would be subject to the errors discussed previously for the angle and position of sections; reliability of this approach would also be affected by the accuracy of identification of the cell structure and the number of points on the grid (see Worked Example 2.4).

Table 2.5 *Properties of electron microscopes and sample preparation methods.*

Property	TEM	SEM
Maximum magnification	1 000 000	300 000
Typical limit of resolution	0.2 nm	1.5 nm
Type of detail discernible	Molecular and macromolecular structure (e.g. molecules, membranes, cytoskeletal elements, chromatin, viruses, etc.). Cross or longitudinal sections.	General and fine surface details of whole cells, tissues and small organisms. Surface detail in 3D image.
Electron beam	Transmitted through stained specimen.	Reflected by specimen coated in gold.
Sample preparation	Classical approach: Specimen is fixed, dehydrated, embedded in resin, sectioned, stained with lead citrate and uranyl acetate, dried and then viewed by TEM. Freeze fracture: Tissue is rapid frozen at 140 °C, freeze-dried, coated with platinum, after which the sample is digested, and the replica viewed by TEM. Direct negative staining: Suspension of fixed or unfixed macromolecules (DNA, viruses and cytoskeletal networks) is adsorbed on to a plastic-coated EM grid and stained briefly with uranyl acetate, dried and viewed by TEM.	Specimens are typically fixed, dried, coated (e.g. with gold), then scanned with a very narrow beam of electrons. Secondary electrons emitted by specimen are detected in a cathode ray tube, creating an image of the specimen surface.

- Once again, measurements on multiple images for several sections would give statistically meaningful data as a single section may not necessarily be representative of the whole specimen.

Worked Example 2.4 Morphometric analysis of electron micrographs

If a grid were superimposed on the image shown schematically in Figure 2.11, and coincidence of grid points (points of intersection of grid lines) with organelles or 'ground cytoplasm' recorded, it is possible to get an estimate of the abundance of each organelle as % total cell volume as follows:
Total number of grid points = 48. Number of 'hits' is expressed as a % of total grid points.

Mitochondria	=	$(3/48) \times 100 = 6.2\%$
Nucleus	=	$(12/48) \times 100 = 25.0\%$
Golgi bodies	=	$(2/48) \times 100 = 4.2\%$
RER	=	$(2/48) \times 100 = 4.2\%$
SER	=	$(3/48) \times 100 = 6.3\%$
Free polyribosomes	=	$(1/48) \times 100 = 2.1\%$
Ground cytoplasm	=	$(25/48) \times 100 = 52.0\%$

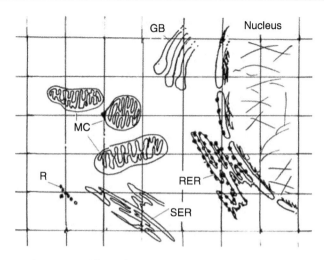

Figure 2.11 *Electron microscope grid morphometry.*

It is possible to estimate the actual size of cell structures from a photographic image, provided that the final magnification is known (see Worked Example 2.5). In this case, the measurement (e.g. in mm) is divided by the magnification factor, which should include the factor for photographic enlargement. The value is then converted into appropriate units.

Worked Example 2.5 Calculation of organelle sizes from electron micrographs

In an image with a final magnification of ×80 000, a structure identified as a mitochondrion was found to be 20 mm in length and 13 mm in width. What are its actual dimensions?

First of all, divide the measured size in the image by the magnification factor to obtain the size in mm. Then, referring to Table 2.1, convert the value into appropriate units.

a. Mitochondrion length = 20 ÷ 80 000 = 0.00025 mm
These are not appropriate units as organelle size is normally expressed in μm or nm.
Length in μm = 0.00025 ÷ 1000 = 0.25 μm
Length in nm = 0.00025 ÷ 10^6 = 250 nm
b. Mitochondrion width = 13 ÷ 80000 = 0.0001626 mm
Length in μm = 0.0001626 ÷ 1000 = 0.16 μm (to 2 decimal places)
Width in nm = 0.0001626 ÷ 10^6 = 163 nm (rounded up to nearest whole number)

Student exercises

In an image with magnification of ×60 000, a mitochondrion was measured at 25 mm in length and 12 mm in width. Calculate its actual dimensions.

2.6 Summary

- Light microscopy has a limit of resolution of approximately 200 nm and a maximum magnification of ×1500, allowing visualization of structures down to the size of mitochondria.
- Electron microscopy can achieve a limit of resolution down to 0.2 nm (TEM) and a maximum magnification of ×10^6, allowing visualization of ultrastructural detail of the cell.
- Bright field microscopy is used to determine cellular organization at the tissue level.
- Contrast enhancement in light microscopy is achieved using chemical stains such as hematxylin/eosin or, in the case of unstained specimens, by the use of contrast enhancement optics in phase contrast or differential interference contrast microscopy.
- The tissue and intracellular distribution of specific molecular components can be observed using antibody staining techniques, although the actual visualization of the molecules at an ultrastructural level requires the use of transmission electron microscopy.
- Contrast in TEM is enhanced by staining with electron-dense heavy metals such as uranyl acetate.
- Interpretation of electron and light micrograph images is aided by developing knowledge of (a) general tissue organization and (b) the appearance and location of organelles within cells.
- When interpreting images, remember that the appearance of a specimen in a thin section is partly dependent on the angle of cut.

3

SPECTROPHOTOMETRY

3.1 Introduction

A popular instrument used by students on biological sciences courses in higher and further education is the spectrophotometer. Proteins, peptides, amino acids, nucleic acids, nucleotides and other cellular metabolites can be measured by the application of spectrophotometry. In this section, we will focus on the wavelengths of the electromagnetic spectrum (200–800 nm), covering the wavelengths of ultraviolet light (200–350 nm) and visible light (350–800 nm).

Many laboratory procedures require the use of a spectrophotometer (a machine that measures the transmission and absorbance of light) and modern spectrophotometers are easy to use. However, to fully understand the results generated and to avoid any errors, it is advisable to have a basic understanding of the machines and of the physical laws involved.

3.2 The Electromagnetic Spectrum

The electromagnetic spectrum (see Figure 3.1) is a collective term given to a broad band of radiations, ranging from the longer wavelengths (see Figure 3.2) of radio waves (wavelengths ≥ 1.0 cm) to the shorter more energetic wavelengths of the gamma rays (wavelengths ≤ 1.0 pm). Photons (hv) are the basic unit of the electromagnetic spectrum at every wavelength, and they have the properties of both a wave and a particle (wave-particle duality).

These properties include the following:

- They have zero mass.
- In free space, they move at the speed of light ($c = 3.0 \times 10^8$ ms^{-1}).
- They are discrete packages of energy, which is related to their wavelength (shorter wavelengths have more energy; see Worked Example 3.1).
- They can be absorbed (destroyed) and created (emitted).[1]
- They can have particle-like interactions (i.e. collisions) with electrons and other atomic particles.

Key to Figure 3.2: λ, (lambda) wavelength is the distance between wave crests; ν, (nu) frequency is the number of wave crests that pass a point in a given amount of time.

Wavelength and frequency are related:

$$C = \lambda v \left(\text{where } C = \text{the speed of light, } 3.0 > 10^8 \text{ ms}^{-1} \right)$$

$$v = c/\lambda$$

Basic Bioscience Laboratory Techniques: A Pocket Guide, Second Edition. Philip L.R. Bonner and Alan J. Hargreaves.
© 2022 John Wiley & Sons Ltd. Published 2022 by John Wiley & Sons Ltd.

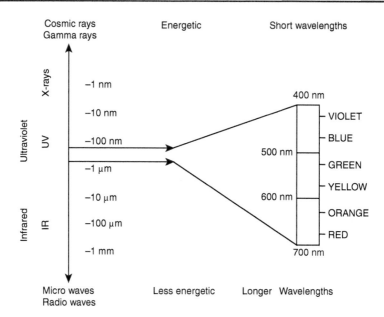

Figure 3.1 *The electromagnetic spectrum.*

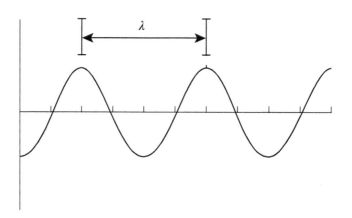

Figure 3.2 *Wavelength and frequency.*

Worked Example 3.1 Different Electromagnetic Wavelengths have Different Energy Levels

Max Planck showed that the energy available at a given wavelength is given by $E = h\nu$ where h, Planck constant (6.63×10^{-34} Js); ν, wave frequency The wave frequency $\nu = c/\lambda$ where c, speed of light (3×10^8 ms^{-1}); λ, wavelength (usually in nm).

- *The energy of a photon at 300 nm*

$$v = c/\lambda$$

$$\frac{3 \times 10^8 \, ms^{-1}}{300 \times 10^{-9} \, m} = 1 \times 10^{15} \, \text{Hertz}$$

$$E = hv$$

$(6.63 \times 10^{-34}) \times (1 \times 10^{15}) = 6.63 \times 10^{-19}$ Joules per photon at 300 nm
- *The energy of a photon at 600 nm*

$$v = c / \lambda$$

$$\frac{3 \times 10^8 \, ms^{-1}}{600 \times 10^{-9} \, m} = 0.5 \times 10^{15} \, \text{Hertz}$$

$$E = hv$$

$6.63 \times 10^{-34} \times 0.5 \times 10^{15} = 3.32 \times 10^{-19}$ Joules per photon at 600 nm. The energy of a photon with a wavelength of 300 nm is double the energy of a photon with a wavelength of 600 nm.
- Conclusion: Photons with shorter wavelengths have more energy.

3.3   The Absorbance of Light

Every atom of every molecule has a defined structure with negatively charged electrons rotating in orbitals around the positively charged nucleus of the atom. Each electron orbital has an energy requirement for occupancy, that is, if an electron receives additional energy, it can no longer rotate in its current orbital and is promoted to a vacant orbital with a higher energy requirement. Photons radiated by the sun across the electromagnetic spectrum have wave-like properties (discrete aliquots of energy) and particle-like properties (the ability to interact with other atomic particles) described as wave-particle duality.

When the wide range of photons from the electromagnetic spectrum encounter the atoms of molecules, some of the photons will have the correct amount of energy to promote electrons to higher energy orbitals. If the energy requirement of the electron is matched by the energy in the photon, the energy is transferred to the electron. The electron now has too much energy to remain in the ground state orbital and is promoted to a higher energy orbital (see Figure 3.3). The photon's energy is acquired by the electron in a molecule, which stays in the higher energy orbital, rotating and vibrating for a brief period of time $(1 \times 10^{-15} \text{s})$ depleting the energy supplied by the photon.

The photon's energy has been *absorbed* in its entirety and the photon at that wavelength will no longer exist. Having used up the energy supplied by the photon, the electron returns to its original orbital in the ground state and the whole process can be repeated by the absorption of another photon. Other photons at different wavelengths will not have the exact energy match for the electron and will pass through the atoms of the material.

3.3.1   Fluorescence

In general, molecules that fluoresce are organic molecules that have associated ring structures, and they have both excitation and emission spectra (which can be determined by a scan; see Figure 3.4). In fluorescence, when a photon is absorbed at the excitation wavelength, the excited electron is elevated to higher energy levels. In the higher energy orbital, the electron exhausts some of the received energy; as a result, it no longer has sufficient energy to remain at the higher level and must return to the ground state. However, to be able to re-enter the ground state the electron must shed the remaining energy by re-emitting a photon ('fluorescence') at a longer wavelength[2] (i.e. with less energy). Fluorescence spectroscopy can be used to increase the sensitivity of assays and of the detection of biological components.

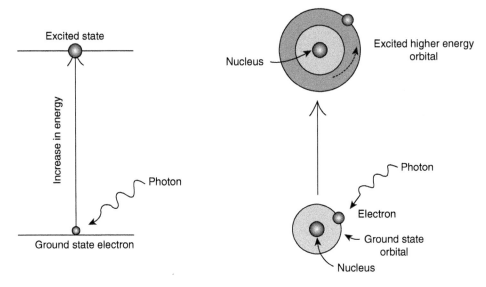

Figure 3.3 *The absorption of a photon.*

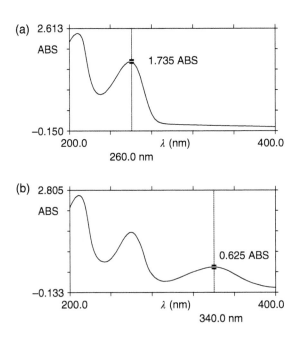

Figure 3.4 *(a) A scan between 200 and 400 nm of 100 mM NAD⁺ using a Beckman 530 spectrophotometer. (b) A scan between 200 and 400 nm of 100 mM NADH using a Beckman 530 spectrophotometer (ABS = absorbance).*

3.3.2 Luminescence

The absorbance of light and fluorescence by molecules depends upon receiving energy from the electro-magnetic spectrum. In contrast, luminescence is the generation of wavelengths of light resulting from a chemical reaction. Chemiluminescence occurs due to a chemical reaction, and bioluminescence occurs as a result of an enzyme-catalyzed reaction. Bioluminescence assays are increasingly popular because of their high sensitivity. Luciferase from fireflies can be used to detect 1×10^{-15} M ATP, and bacterial luciferase can be used to detect 1×10^{-12} M NADH, NADPH and $FMNH_2$.

3.4 Absorption Spectrophotometry

The visible wavelengths of the electromagnetic spectrum are composed of the colours of the rainbow, and the colour of any material is the result of the absorbance of different wavelengths of light, for example, plants appear green in colour (500–600 nm) because they contain pigments that absorb the blue wave-lengths of light between 400 and 500 nm and the red light wavelengths between 600 and 700 nm. The wavelengths that are not absorbed are the wavelengths we can see. In the case of plants, the remaining wavelengths of light (500–600 nm) appear green. The colour of any compound depends upon its structure.

KEY POINTS TO REMEMBER 3.1

The colour of a material or solution is due to the wavelengths of light that have not been absorbed. For example, a solution of copper sulphate is blue because, as white light passes through the liquid, the wavelengths of light which make up the colours red, green and yellow are absorbed leaving the blue wavelengths

---

The wavelengths of light that pass through a solution can be measured by a photosensitive device, and this forms the basis of spectrophotometry. Spectrophotometers (see Section 3.7) can measure the absorb-ance of a solution at all the different wavelengths of UV/visible light and plot these measurements to pro-vide a 'scan'. The scan (see Figure 3.4a, b) shows that compounds can absorb photons of light across a wide range of the UV/visible part of the EM spectrum. However, there is usually a wavelength of light at which the compound will absorb the most photons. This maximally absorbing wavelength is called the lambda max (λ_{max}, e.g. λ_{410nm}) and this is characteristic of the compound absorbing the light.

KEY POINTS TO REMEMBER 3.2

Changes in the structure of molecules can subtly alter their absorbance pattern, for example, nicotinamide adenine dinucleotide (NAD⁺) (Figure 3.4a) and reduced nicotinamide adenine dinucleotide (NADH) (Figure 3.4b) are vital cellular cofactors. Both molecules absorb light at 260 nm (the characteristic wavelength where nucleotides absorb maximally. NAD⁺ and NADH contain adenine a nucleotide), but only the reduced cofactor (NADH) also absorbs light at 340 nm. This dramatic wavelength change can be used to follow the course of chemical reactions catalyzed by dehydrogenase enzymes (enzymes which require the cofactor NAD⁺/NADH). The appearance or disappearance of NADH in the dehydrogenase assays can be monitored by UV spectrophotometers that can record the absorbance of light at 340 nm.

3.5 The Laws Governing the Absorbance of Light

The transmission of light is measured by the percentage of the incident light that passes through the sample:

$$\%\text{Transmittance } (T) = (I / I_0) \times 100\%$$

where I, intensity of transmitted light; I_0, intensity of incident light.

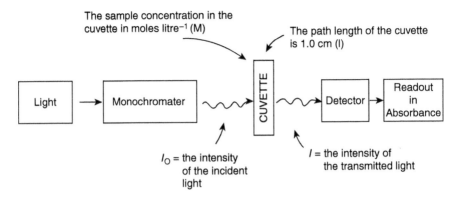

Figure 3.5 *Incident and transmitted light.*

In this case, the light intensity is the number of photons interacting second^{-1}

When % Transmittance $(T) = 100\%$ a solution is totally transparent and when % Transmittance $(T) = 0\%$ a solution is opaque.

If $T = 40\%$, then 40% of the photons passing through the sample reach the detector and the other 60% are absorbed by the sample.

Absorbance (see Figure 3.5) is the $\log_{10}$ of the incident light (I_0) divided by the transmitted light (I).

$$A = \log_{10}(I_0/I) = \log_{10}(1/T)$$

Absorbance (a logarithmic value) has no units and takes values between zero (0) and infinity (∞)

- If $A = 0$, then no photons are absorbed.
- If $A = 1.00$, then 90% of the photons are absorbed; only 10% of the photons reach the detector $\%T = 10\%$.
- If $A = 2.00$, then 99% of the photons are absorbed; only 1% of the photons reach the detector $\%T = 1\%$.
- Most spectrophotometers will be able to display both the absorbance of a sample and/or the transmitted light through the sample.

3.5.1 Attenuance of Light (Optical Density)

Absorbance and light scattering (attenuance or optical density) are different phenomena. The absorbance of light by a liquid at a specific wavelength is a measure of the photons captured by a transparent solution. When solutions containing cells, plastids and particles are placed in a spectrophotometer they may absorb light, but they will additionally also deflect light from the detector resulting in a decreased signal. This reduction in measured light given by these cloudy solutions is called the attenuance of light. In microbiology, the term optical density (OD) is used.

KEY POINTS TO REMEMBER 3.3

The absorbance of light is not the equivalent of the attenuance (optical density) of light. When the particulate containing solution is sufficiently diluted, attenuance will approximate to absorbance.

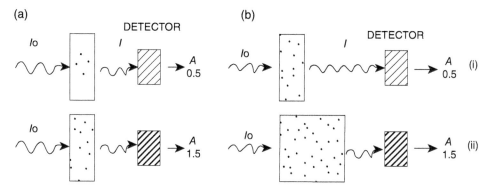

Figure 3.6 *Beer–Lambert Law is related to (a) the concentration of the sample (the concentration of the cuvette (ii) is three times the concentration in cuvette (i). This is reflected in the increase in absorbance) and (b) the path length that the light must travel through (the concentration of the solution in (i) is the same as in (ii) but the path length in cuvette (ii) is three times longer than the path length in cuvette (i). This is reflected in the increase in absorbance.*

3.6   The Beer–Lambert Law

Most scientists choose to report data in terms of absorbance rather than transmittance, because absorbance is related to the concentration of the solution. In a spectrophotometer, the absorbance readout is a measure of the number of photons absorbed by a transparent solution. It follows that if there are more molecules in the solution (i.e. at a higher concentration), the absorbance will be greater (see Figure 3.6a). This relationship is linear, in that, as the concentration of a solution increases, so does the absorbance.

In a similar manner, the longer the path length that light must travel through a solution before reaching the detector, the greater the chance a photon will be absorbed. Therefore, absorbance is also related to the path length the light must travel through the solution (see Figure 3.6b).

The Beer–Lambert law states that the absorbance of a solution is a function of three factors:

- The path length (L) of the light must pass through the cuvette holding the solution (this has been standardized in most laboratory instruments to be 1.0 cm). The wells of a 96 well microplate will have a different pathlength for different volumes. The manufacturers will supply the relevant pathlength).
- The concentration (C) of the reagent in molarity (M).
- The absorptivity coefficient (ε), which is characteristic of the analyte in solution (ε is called the molar absorptivity coefficient (M^{-1} cm^{-1}) if the concentration is measured in molar (M); see Chapter 1).[3]

$$\text{The Beer–Lambert Law: Absorbance } (A) = L\varepsilon C$$

where ε, molar absorptivity coefficient (M^{-1} cm^{-1}); C, molar concentration of solute (M); L, the path length of sample cell/cuvette or microplate well (cm).

Absorbance displays a simple dependence on the concentration of a solution and the path length the light must travel through (The Beer–Lambert Law). This relationship allows scientists to convert an absorbance value from a spectrophotometer into a concentration, thus making the change in absorbance a quantifiable value

---

KEY POINTS TO REMEMBER 3.4

The molar absorptivity coefficient (ε) is valid only for the specified wavelength. For the molar absorptivity (ε), many different compounds may be obtained from reference books, and this represents the absorbance of a 1.0 M solution at a *defined* wavelength in a 1.0 cm cell/cuvette path length at 25 °C. For example, the molar absorptivity coefficient (ε) of NADH at 340 nm is **6.22×10^3** (M^{-1} cm^{-1}) at 25 °C.

Spectrophotometers measure the absorbance of solutions between 0 and 3.0 (see Section 3.5). This means that a 1.0 M solution of NADH will need to be diluted to bring it onto the scale of the machine.

- A 1.0 M solution of NADH at 340 nm would have an absorbance of 6220.
- 1/10 dilution results in a 100 mM solution of NADH with an absorbance of 622.
- 1/100 dilution results in a 10 mM solution of NADH with an absorbance of 62.2.
- 1/1000 dilution results in a 1 mM solution of NADH with an absorbance of 6.22.
- 1/10 000 dilution results in a 100 μM solution of NADH with an absorbance of 0.622.

A spectrophotometer will be able to provide an absorbance reading when the concentration of NADH is between 2 and 400 μM.

3.6.1   Limitations of the Beer–Lambert Law

The Beer–Lambert Law is a linear relationship between the absorbance and the concentration up to 1.0 M. A plot of concentration against absorbance may deviate from linearity at high concentration values forming a plateau, possibly due to interactions between molecules. The Beer–Lambert Law will not apply in the plateau area of the plot and readings should only be taken in the *linear* part of the graph, usually at low concentrations.

The Beer–Lambert law can be used to determine the concentration of reagents in solution providing a molar absorptivity coefficient (ε) is available (see Worked Example 3.2) and to quantify enzyme catalyzed reactions whether the substrate is consumed or the product appears (see Worked Example 3.4).

Worked Example 3.2

The molar absorptivity coefficient (ε) of NADH at 340 nm is **6.22×10^3** (M^{-1} cm^{-1}) at 25 °C, and the path length of the cell is **1.0** cm. The relative molecular mass (Mr) of NADH is 663. (Please note that the wavelength number and temperature are quoted for reference; they are not components of the Beer–Lambert Law equation.)

Question 1: The absorbance of a solution of NADH measured at 340 nm is 1.12 in a cuvette with a pathlength of 1.0 cm.

Calculate the concentration of the NADH solution.

Using the Beer–Lambert Law: Absorbance $(A) = L\,\varepsilon\,C$

$$1.12 = 1.0 \times 6220 \times C$$

$$C = \frac{1.12}{6220}$$

$$C = 0.00018\,M$$

Answer: 0.00018 M or 0.18 mM or 180 μM (see Chapter 1 for information on the concentration of solutions).

Question 2: The absorbance of a solution of NADH measured at 340 nm is 0.34. Calculate the concentration of the NADH solution.

Using the Beer–Lambert Law: Absorbance $(A) = L\,\varepsilon\,C$

$$0.34 = 1.0 \times 6220 \times C$$

$$C = \frac{0.34}{6220}$$

$$C = 0.000054\,M$$

Answer: 0.000054 M or 0.054 mM or 54 μM (see Chapter 1)

3.7 Spectrophotometers

3.7.1 Introduction

A variety of instruments have been manufactured to measure the absorbance of light by liquids.

Colorimeter (Colourimeter) (instruments that measure coloured reagents) are relatively unsophisticated machines that use:

- prisms to refract white light into its constituent colours,
- filters to abstract interfering wavelengths of light,
- a cuvette to hold the sample and galvanometers to record the intensity of the transmitted light.

Spectrophotometers (see Figure 3.7) are instruments which usually consist of:

- a light source (sometimes two; UV and visible light sources),
- a monochromator (a device used to generate light at different wavelengths),
- a sample chamber with lid, containing a cuvette to hold the sample (possibly temperature controlled; this can be used to monitor the progress of enzyme-catalyzed reactions at 37 °C),
- a photomultiplier to detect the transmitted light, and
- a data analysis software package.

In most spectrophotometers, the light absorbed by the solution the sample is dissolved in (the background or reference) is subtracted from absorbance of the sample. In single beam instruments, this is achieved by measuring the absorbance of reference solution in a cuvette (sometimes called 'the blank') and setting the absorbance to zero before measuring the absorbance of the sample at the same wavelength. The zero must be reset if the wavelength is changed. Dual beam spectrophotometers have two cuvette holders: one for the reference solution and one for the sample. Light at the required wavelength is directed intermittently on the reference and sample cuvettes by means of a rotating mirror. In this configuration, the absorbance of the sample can be constantly compared with the reference solution, which is helpful if the wavelength is changed. A pair of optically matched cuvettes may be needed for accurate measurements.

Diode array spectrophotometers have a different optical arrangement to conventional spectrophotometers. The sample chamber is open as light is monochromated by a fixed angle diffraction grating (see below) after leaving the sample. The entire spectrum of monochromated light is detected by an array of photosensitive diodes which means no wavelength changing mechanism is required. However, wavelength resolution may be limited to 2 nm increments using these diodes.

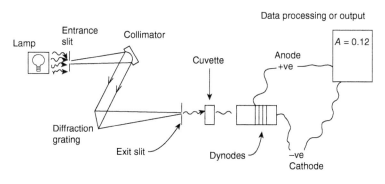

Figure 3.7 *A simple spectrophotometer.*

3.7.2   The Light Source

To cover the complete range of the UV/light area of the electromagnetic spectrum (200–800 nm), two different light sources are required. A deuterium arc lamp provides UV light from 200 to 360 nm and a tungsten filament lamp, which provides light from 360 to 800 nm. The outputs of these bulbs complement each other as there is little overlap in their wavelength ranges. In general, the deuterium lamps are more expensive to replace and have a shorter life span than the tungsten bulbs. Remembering to switch the deuterium bulbs off after the instrument has been used will lower the running costs. Xeon flash lamps generate a wide range of wavelengths covering both the UV and visible ranges, and they can reduce the cost because they are only used when the absorbance of the sample is being measured.

3.7.3   The Optics Used to Generate Wavelengths of Light

White light from the light source is first collimated (see Figure 3.7) into a parallel beam by a curved mirror before being refracted by a prism into its constituent colours. However, a drawback with prisms is that the shorter wavelengths of light (i.e. blue wavelengths) are dispersed more than the longer wavelengths of red light due to the density of the prism. A diffraction grating[4] consists of a silica disc with many finely scored lines (≥2000 mm^{-1}) etched onto its surface. The grating diffracts light into many parallel beams travelling in different directions. The direction of the beams depends on the wavelength of light and spacing of the lines on the diffraction grating. The advantage of diffraction gratings is that they can resolve all the wavelengths of light (from blue to red) independent of the wavelength number. The diffraction grating can then be angled to ensure that the selected wavelength of light impinges on the sample to be measured.

3.7.4   The Slit Width

The light leaving the diffraction grating exits through a slit before encountering the sample in the cuvette and ultimately the detector. If the width of the slit is increased beyond 1.0 nm, this will result in a decrease in the observed absorbance. Most spectrophotometers with diffraction gratings will have a fixed exit slit width to provide a controlled spectral band width.

3.7.5   The Cuvette

The device to hold the sample in the light beam of the colorimeter or spectrophotometer is called a cell or a cuvette (see Figure 3.8). When light is measured in the visible wavelengths of the electromagnetic spectrum, the cuvette can be made from glass, quartz or plastic. But, because glass absorbs light in the UV wavelengths of the electromagnetic spectrum, cuvettes made of quartz or special plastics must be used when measuring UV wavelengths (200–360 nm). In practice, the absorbance properties of the plastic cuvettes allow measurements above 250 nm, which is sufficient to measure both nucleotides/nucleic acids (260 nm) and proteins/peptides (280 nm). Optically matched pairs of cuvettes should be used in dual beam instruments.

The cuvette is the device that provides the interface between the sample and the spectrophotometer. This means that, if it is used incorrectly, it could be a major source of inaccurate measurements.

Cuvettes have a clear face and an opaque face (see Figure 3.8). They may also have a mark to indicate the front and the back of the clear face. The cuvette is placed into a sample holder to ensure reproducible alignment with the light beam. Then the lid should be closed to prevent stray light from interfering with the readings.

- In teaching laboratories, the two most common volumes for cuvettes that are used with bench top spectrophotometers are 1.0 and 3.0 ml with a 1.0 cm path length. Other cuvettes are available for more specialized applications.

Mark to indicate the
front of the cuvette

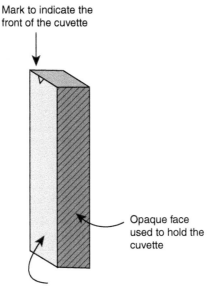

Opaque face
used to hold the
cuvette

Transparent face to be placed
in the light beam of the spectrophotometer

Figure 3.8 A diagram of a spectrophotometer cell/cuvette.

- Always hold the cuvette by the opaque face and place the clear face into the path of the spectrophotometer light beam with the mark (see Figure 3.8) facing the light beam.
- Older cuvettes may have imperfections (small scratches). This is not usually a problem if the same cuvette is used for both the reference solution (blank) and the sample solution. In addition, the cuvette should always be presented to the light beam in the spectrophotometer in the same orientation for each measurement.
- Check that the transparent surfaces are clean. If required wipe the surface with a dry piece of tissue paper before use.
- Most spectrophotometers will accurately measure the absorbance between 0 and 2.0 (sometimes 0–3.0; see Section 3.5). Concentrated samples with an absorbance outside this range will register *off scale* (show a flashing or continuous reading of 2.0 and above). When you place the sample liquid in the cuvette, check to see if you can see through the sample. If you cannot see through the sample, the spectrophotometer will also struggle to read the absorbance.
- Check to see if there are any floating particles in the sample. These floating particles may deflect the light preventing the light from reaching the detector (attenuance) giving an inappropriate reading. In addition, the particles may move through the light beam on convection currents generated in the cuvette by the heat of the machine. In this case, the readings will rise and fall as the particles move through the light beam. These particles will attenuate the light beam and be a source of error. If there are particles in the sample, remove the liquid and filter or centrifuge the sample (see Chapter 5) to remove the particles before replacing the sample in the spectrophotometer.
- Atmospheric gases are more soluble in cold liquids and will be released as the temperature rises. If cold reagents are placed into the spectrophotometer, the released gases may form bubbles that may pass through the light beam causing fluctuations in the reading. If this is suspected, take the cuvette out and gently tap the cuvette on the bench to release the bubbles from the inner surface of the cuvette. Allow the reagents to warm up to operating temperature and then replace the cuvette in the spectrophotometer.
- In addition, cold reagents placed in a warm machine may result in condensation on the cuvette's surface generating anomalous results. If this happens, remove the cuvette, dry the surface with a tissue and return the sample to the machine.

- More sophisticated machines will have temperature-controlled cuvette holders.
- Many laboratories will have 'nano drop' spectrophotometers which can measure the concentration of nucleic acids (260 nm), proteins and peptides (280 nm) in small sample volumes (0.5–2.0 μl).

KEY POINTS TO REMEMBER 3.5

Glass absorbs light in the UV region. Quartz must be used for prisms, gratings and cuvettes (disposable plastic cuvettes are also available to measure wavelengths above 250 nm) when measuring the absorbance of a liquid in the UV region of the EM spectrum.

3.7.6 The Detector

The light that has not been absorbed by the sample solution passes through the cuvette and travels to the detector. Photomultipliers are routinely used in spectrophotometers as detectors. The photons of light strike a photocathode which releases electrons in proportion to the photons striking the surface. The released electrodes strike dynodes set up in a series of increasing potential. When the electrons from the photocathode strike the first dynode, it causes a secondary release of electrons, which is enhanced with each subsequent dynode. The electron density reaching the anode from the final dynode will be multiplied but remains proportional to the original electron emission from the photocathode. The light intensity is reflected by the amount of current generated, and this signal is processed for viewing on the readout of the spectrophotometer.

3.7.7 The Use of High-throughput Microplate Readers

The use of spectrophometry is a valuable technique because the change in absorbance provides quantifiable data, making it a core technique for many academic and industrial scientists. The use of 1.0 or 3.0 ml cuvettes is appropriate in many situations (see Table 3.1), but many assays can be scaled down in volume to be accommodated into a 96-well plastic plate format (see Figure 3.9a). These 96-well plastic plates (12 columns + 8 rows; typical well column volume of 300 μl) were originally designed to be used in antibody-based assays, such as enzyme-linked immunosorbent assays (ELISA) (see Chapter 9), but they have also been adapted to work in a wide variety of enzyme assays. The 96-well format allows a reduction in the reagents and enzymes used, while at the same time allowing numerous measurements to be recorded. Spectrophotometers adapted to the 96-well microplate format provide rapid analysis of the assays performed (see Figure 3.9b).

Table 3.1 *Some popular colourimetric assays to measure the concentration of biological molecules.*

Detection reagent	Target analyte	Detection wavelength (nm)
Ninhydrin	Amino acids	570
Coomassie blue	Protein	595
Bicinchoninic acid (BCA)	Proteins and peptides	572
Diphenylamine	DNA	595
Orcinol (Bial's reagent)	RNA	665
Ammonium molybdate	Inorganic phosphate	600
Folin's phenol reagent	Tyrosine, phenols	660
5,5'Dithiobis-(2-nitrobenzoic) acid	–SH (a sulphur-containing alcohol analogue)	412
3,5-Dinitrosalicylic acid	Reducing sugars	480, 570

(a)

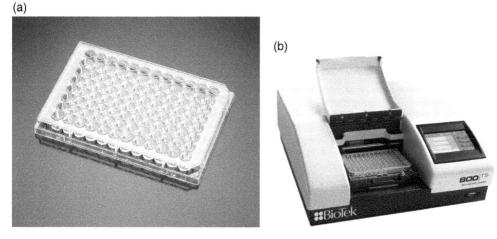

(b)

Figure 3.9 *(a) A Corning Falcon 96-well microplate (taken from https://ecatalog.corning.com/life-sciences/ b2c/UK/en/Microplates/Assay-Microplates/96-Well-Microplates/Falcon%C2%AE-96-well-Polystyrene-Microplates/p/353072). (b) A 96-well microplate reader. (This model is from Biotek which is part of the Agilent group of companies. There are many different examples of 96-well microplate readers.) Taken from https:// www.biotek.com/products/detection-microplate-readers/800-ts-absorbance-reader/*

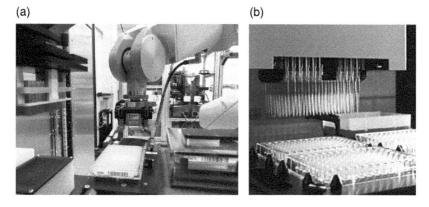

Figure 3.10 *(a) A high-throughput robot moving a multi-well microplate. (b) Robotic pipetting of reagents into high-throughput screening microplates. (a) taken from https://www.ncardia.com/innovations/automated-hts-drug-discovery.html. (b) taken from https://www.worldofchemicals.com/15/chemistry-articles/rapid-lead-compounds-discovery-through-high-throughput-screening.html. Taken from Sigma Aldrich.*

3.7.7.1 High-throughput Screening (HTS)

Pharmaceutical companies use the multi-well plastic plate format of enzyme assays when they are screening for 'lead' (leading) compounds in their design of new pharmaceutical agents. To reduce the costs and to increase the throughput of the reagents screened, the well numbers in the plastic plates increases from 96 to 192, 384, 1536, 3456 and 6144 (all multiples of 96). The sheer number of reagents to be screened and the small volumes of reagents used (typically in the nl range with 6144 well plate format) lend this work to the use of robotics (see Figure 3.10a, b). The chemical reagents to be screened are dissolved in a solvent such as dimethyl sulphoxide (DMSO) and can be aliquoted into the library plates. The robotic arms can transfer small volumes of these compounds into the plates containing the necessary reagents for an enzyme assay.

The reaction is initiated by the addition of an accurate injection of fresh enzyme or enzyme substrate into the wells of a multi-well plate. Following a suitable time period, the reaction is stopped, and results compared to an assay in the presence of potential inhibitors. The robotic screening is left to run automatically until all the chemicals in the company's chemical compound library have been assessed and the large amount of data generated is processed by a data management system. A positive result may lead to the pharmaceutical company synthesizing new reagents with a similar structure, which will also be tested in screening process until an effective and non-toxic compound is identified.

Things to remember when performing microplate assays

- 96-well plastic microplates come in a number of different well shapes (e.g. flat, round or conical bottoms), and they also vary in the chemistry of the plastic used (e.g. activated and high binding plastics). Make sure you are using the appropriate format for the required assay and endeavour to use the same format in subsequent assays.
- As a rough guide, a 100 µl volume in a well of a 96-well microplate will have a 0.28 cm pathlength, 200 µl 0.56 cm and 250 µl a 0.7 cm pathlength.
- The different shapes of the wells in a microplate will alter the length (cm) of the light beam. Remember to enter the correct light beam path length (this can be found at the microplate manufacturers web site) when quantifying the results using the Beer–Lambert law (see Worked Example 3.5).
- Different buffers and the concentration of some reagents will alter the meniscus of the liquid in the microplate which will also alter the pathlength of light in a microplate well. Fortunately, many microplate readers will have compensation factors built into the software.
- If you are concerned that you are generating anomalous readings with the microplate reader, compare the result obtained using a microplate volume with the result obtained for a 1.0 ml volume using 1.0 ml (or 3.0 ml) cuvette with a 1.0 cm pathlength. Then use the Beer–Lambert Law to factor in an adjustment.
- The light beam from the microplate spectrophotometer goes through the plate wells from top to bottom or vice versa. Any stain or fingerprint on the bottom of the wells will divert or absorb the light from the light beam producing an incorrect value. Always hold the 96-well plate by the edges and always label the plate on the edges.
- In most assays, the length of time the enzyme is in contact with the substrate is critical. To ensure that the assay runs for a similar length of time, use an 8-channel pipette to start the reactions at the same time in each well of one column in the 96-well plate (see Figure 3.11). Alternatively, if a single channel pipette is used, stagger the addition of the reaction initiating reagent (e.g. enzyme) to each individual well by 10–15 s.
- A source of error in using microplates are the wells around the edge of the plate (rows A and H and columns 1 and 12). It is thought that the evaporation rate is greater in these wells, and this source of error can be reduced by using a plastic 96-well microplate lid, sealing the plate with a microtiter plate adhesive sealing film or by using both the film and a lid. Alternatively, avoid the wells at the edge of the plate for crucial experiments.
- One of the great values of using microplates is the number of replicates that can be included. The more replicates used to calculate the mean, the smaller is the deviation about the mean (see Chapter 4, Section 4.2.3).
- Pipetting accuracy will be revealed by the value of the deviation about the mean (see Chapter 4, Section 4.1).
- Dilution precision will be revealed by the R^2 value in the linear regression analysis (see Chapter 4, Section 4.1).

3.8 Applications of Spectrophotometry in Bioscience

3.8.1 Direct Measurements of Biological Molecules

As mentioned in the introduction, spectrophotometers can be used to measure the concentration of many common biological molecules. Some of the macromolecular structures such as proteins and nucleic acids can be measured directly. Proteins characteristically absorb light at 280 nm due to the presence of two

Figure 3.11 *A variable volume 8 channel micropipette. Source: Thermo Fisher Scientific Inc.*

amino acids (tryptophan and tyrosine) with aromatic rings as the functional groups in their structures. Nucleic acids characteristically absorb light at 260 nm due to the presence of nucleotides (cytosine, adenine, thymine, uracil and guanine) within their structures.

For a quick method to measure the concentration of proteins in solution place the solution in a spectrophotometer and measure the absorbance at 280 nm. For a pure protein solution, a 1% (w/v) absorptivity (ε 1%) value at 280 nm can be determined and the acquired reading gives a direct measure of the protein's concentration. This method has problems when used with samples derived from homogenized samples (see Chapter 5, Section 5.3.2). The crude homogenates will also contain metabolites, peptides and amino acids which will also absorb light at 280 nm. The interference may be reduced by dialysis (see Chapter 7, Section 7.4.1) to remove small molecular weight metabolites before reading the sample's absorbance in the spectrophotometer. In addition, a correction factor can be used if nucleic acids are still present in the protein solution.

$$\text{Protein}\left(\text{mg ml}^{-1}\right) = 1.55\left(A_{280\,\text{nm}}\right) - 0.76\left(A_{260\,\text{nm}}\right)$$

Alternatively, a standard calibration graph (see Chapter 4) of known concentrations of bovine serum albumin (x-axis) can be plotted against the absorbance at 280 nm (y-axis). The absorbance values of samples can be read from the standard graph and converted into a protein concentration. The problem with this method is that not all proteins contain the same number of tyrosine and tryptophan residues as bovine serum albumin and the concentration of some proteins will be over- or under-estimated.

The purity of a sample of DNA can also be determined by measuring the absorbance at both 260 and 280 nm. A ratio between 1.7 and 2.0 represents good-quality DNA. If the sample is turbid an additional correction factor can be included by also measuring the absorbance at 320 nm.

$$\text{DNA purity} \left(260 / 280\,\text{nm} \right) = \frac{\left(A_{260\,\text{nm}} - A_{320\,\text{nm}} \right)}{\left(A_{280\,\text{nm}} - A_{320\,\text{nm}} \right)}$$

Good-quality single-stranded RNA produces a 260/280 nm ratio of 1.9–2.1 in 10 mM Tris buffer pH 7.5 (the pH influences this ratio).

3.8.2 Chromophore Assays

Very few metabolites in biology absorb light in the visible range of the electromagnetic spectrum. This could be problematic if a spectrophotometer with only a tungsten light source for measuring visible light is available. Over the years, methods have been developed to attach reagents to target molecules to produce a coloured reaction product, which can then be measured in the visible light range of the electromagnetic spectrum.[5] Ideally, the reagent should react with a limited group of compounds to provide selectivity and form the basis of a quantitative assay (see Table 3.1). If the molar absorptivity coefficient (ε) of the coloured product is known, this can be used to determine the concentration of the metabolite in the sample (see Section 3.6). Alternatively, a standard calibration graph can be prepared (see Worked Example 3.3 and data analysis Chapter 4).

Using this approach, there can be drawbacks measuring samples derived from crude cell homogenates. Metabolites may be present that will also react or partially react with the reagent, making it difficult to determine accurate results.

Appropriate sample preparation may be required to remove contaminating molecules to increase the accuracy of the results.

Worked Example 3.3 To Determine the Free Amino Acid Concentration in a Crude Cell Homogenate

The tissue should be homogenized and clarified (see Chapter 5, Section 5.3). The addition of ice-cold trichloroacetic acid (TCA) to a final concentration of 7% (w/v) will precipitate any proteins in the supernatant. This precipitate can be removed by centrifugation at $13\,000 \times g$ for 30 min at 4 °C. The protein-free supernatant will contain small molecular weight metabolites including the free amino acids.

- Prepare a solution of alanine for the standard calibration graph in 7% (w/v) TCA.
- Prepare a reference (blank). For this example, it would be prudent to use 1.0 ml of 7% (w/v) TCA as the reference blank.
- Prepare 1.0 ml dilutions of the standard alanine in triplicate.
- Prepare 1.0 ml dilutions of the sample in triplicate.
- Add 0.5 ml 0.1% (w/v) ninhydrin and boil for 5 minutes.
- An intense blue colour will develop.
- Cool the standards and samples to room temperature and set the wavelength number on the spectrophotometer to 570 nm.
- Use the reference sample to set the absorbance to zero.
- Measure the absorbance of the standards and samples.
- Prepare a calibration graph (see Chapter 4) by plotting the known concentration of standard alanine (x-axis) against the recorded absorbance for each standard (y-axis).
- Use the calibration graph to convert the absorbance of the diluted sample into an amino acid concentration. At least one of the dilutions should fall on the linear part of the calibration graph and remember to multiply up by the dilution factor (see Chapter 4).

Note: At high concentrations, the graph may deviate from linearity and form a plateau. This may be because the reagent is not present at a high enough concentration to produce enough colour or it may be due to the formation of interactions between molecules. Either way, the Beer–Lambert Law is only valid on the linear part of the graph.

If the absorbance of the sample does not fall on the linear part of the graph, the sample should be further diluted, and the reaction repeated until the absorbance reading does fall within the linear part of the calibration graph.

3.8.3 Enzyme Catalyzed Reaction

Spectrophotometry can be used to follow the course of an enzyme catalyzed reaction. For example, dehydrogenase enzymes use the cofactors NAD$^+$/NADH. Thus, the appearance or disappearance of NADH can be quantified by following the course of the reaction at 340 nm using a 1.0 ml cuvette with a 1.0 cm light pathlength (see Worked Example 3.4 and Figure 3.4b). For convenience and to reduce the use of reagents, the course of enzyme catalyzed reactions can be followed using a 96-well microplate; for example, when measuring the activity of the peptidase trypsin using a substrate with a small molecular mass, e.g. benzoyl-L-arginine-4-nitroanilide (see Chapter 3, Section 3.7.7. and Worked Example 3.5).

Alternatively, the activity of a peptidase (e.g. trypsin), which can cleave internal peptide bonds, can be followed using a chemically modified protein such as azocasein. If the molar mass of the protein substrate is not known, a mass absorptivity coefficient (ε $^{1\% \, (w/v)}$) rather than a molar absorptivity coefficient can be used in the Beer–Lambert law to quantify the peptidase catalyzed reaction (see Worked Example 3.6).

Worked Example 3.4 Lactate Dehydrogenase Assay in a 1.0 ml Cuvette

Pyruvate   Lactate

The enzyme lactate dehydrogenase catalyzes the conversion pyruvate to lactate; as part of the reaction, the cofactor NADH is oxidized to NAD$^+$. The progress of the reaction (the rate of disappearance of NADH) can be monitored by recording the change in absorbance at 340 nm ($\Delta A_{340\,nm}$) in a spectrophotometer.

Into a 1.0 ml cuvette was placed 0.9 ml of 200 mM Tris/HCl buffer pH 7.4, 0.05 ml of 6.6 mM NADH and 0.05 ml of 30 mM sodium pyruvate. The cuvette was placed into a spectrophotometer set to record the absorbance at 340 nm. The non-enzymic background change in absorbance was measured over a five-minute period.

The cuvette was emptied, rinsed and refilled with 0.8 ml of 200 mM Tris/HCl buffer pH 7.4, 0.05 ml of 6.6 mM NADH, 0.05 ml of 30 mM sodium pyruvate and the reaction started by the addition of 0.1 ml homogenized and clarified (see Chapter 5) bovine heart muscle (diluted to 0.5 mg ml^{-1} prior to use in 200 mM Tris/HCl buffer pH 7.4).

The following data were obtained:

Time (min)	0.0	1.0	2.0	3.0	4.0	5.0
Absorbance 340 nm	0.8	0.68	0.56	0.48	0.41	0.35

- Remember that you are measuring a rate (the change in absorbance at 340 nm min^{-1}), so it does not matter too much where the initial absorbance reading starts from (initial A340 nm between 0.4 and 1.2 is fine).
- If the absorbance reading changes very quickly over a 1.0-min period, repeat the experiment with an enzyme preparation that has been diluted.
- Some spectrophotometers will determine the rate of reaction and present the data in a table. Otherwise a graph of A$_{340\,nm}$ against time can be plotted to determine the gradient of the graph that is ΔA$_{340\,nm}$ min^{-1} (remember, if the graph starts to plateau, you must use the initial *linear* part of the graph to determine the rate of reaction over a sensible time period).

The molar absorptivity coefficient (ε) of NADH at 340 nm is 6.22×10^3 (M^{-1} cm^{-1}) at 25 °C, and the path length of the cuvette is 1.0 cm.

The change in absorbance at 340 nm over a one-minute period between, for example, 1.0 and 2.0 minutes) is 0.12.

$$\text{Using the Beer–Lambert Law: Absorbance}(A) = L\varepsilon C$$

$$0.12 = 1.0 \times 6220 \times C$$

$$C = 0.000019 \text{ M} \left(\text{min}^{-1}\right)$$

$$= 19 \ \mu M \left(\text{min}^{-1}\right)$$

$$= 19 \ \mu\text{mol in 1000 ml} \left(\text{min}^{-1}\right)$$

The Beer–Lambert Law provides the concentration in mol l^{-1} (see Chapter 1).

The cuvettes in bench top spectrophotometers have either a 1.0 or 3.0 ml volume.

The amount of product in the 1.0 ml cuvette can be found by dividing by 1000 to give the amount (mol) in 1.0 ml

$$= \frac{19}{1000} \mu\text{mol min}^{-1}$$

$$= 0.019 \mu\text{mol min}^{-1}$$

$$= 19 \text{ nmol of product formed min}^{-1} \text{ in the 1.0 ml cuvette}$$

$$\text{or} \quad = 0.019 \text{ units of enzyme activity in the cuvette}$$

International units of enzyme activity:
1 unit of enzyme activity = 1 μmol of product formed (or substrate consumed) min^{-1}.

There are 0.019 units of lactate dehydrogenase in the 0.1 ml of the bovine heart extract or 0.19 units in 1.0 ml of bovine heart extract (0.5 mg ml^{-1}).

$$\text{Specific activity} \left(\text{units mg}^{-1} \text{ protein}\right) = \frac{\text{units ml}^{-1}}{\text{protein} \left(\text{mg ml}^{-1}\right)}$$

$$= \frac{0.19}{0.5}$$

In this experiment, there are 0.38 units of lactate dehydrogenase activity mg^{-1} of bovine heart protein.

Worked Example 3.5 A Trypsin Assay in a 96-well Plastic Microplate Using Benzoyl-L-Arginine-4-Nitroanalide (formally known as BAPNA or BApNA; benzoyl-L-arginine-para-nitroanalide) (see Figure 3.12)

In a microplate well (total reaction volume 0.2 ml = 0.56 cm light pathlength for Corning Falcon 96-well plastic microplates; see Figure 3.9a), a 10 μl solution of trypsin (0.05 mg ml^{-1}) produced an absorbance change at 405 nm of 0.54 in 5 min. (The molar absorptivity coefficient (ε) of 4-nitroanaline at 400 nm = 12 000, at 405 nm = 9500 and at 410 nm = 8800 M^{-1} cm^{-1}.)

The enzyme solution volume was 25.0 ml and the protein concentration 0.05 mg ml^{-1}.

Questions for the worked example:

a. How many international units of enzyme activity was present?
b. What was the total protein (mg)?
c. What was the specific activity (units mg^{-1})?
d. What was the total activity (units)?

Answers:
a. Absorbance change min^{-1} = 0.54/5
$$= 0.108 \, \text{min}^{-1}$$
The Beer–Lambert law: $A = L \, \varepsilon \, C$

$$0.108 = 0.56 \times 9500 \times C$$

$$\frac{0.108}{0.56 \times 9500} = C$$

$$C = 0.0000203 \, \text{M} \left(\text{min}^{-1} \right)$$
$$= 20.3 \, \mu\text{M} \left(\text{min}^{-1} \right)$$
$$= 20.3 \, \mu\text{mol in 1000 ml} \left(\text{min}^{-1} \right)$$

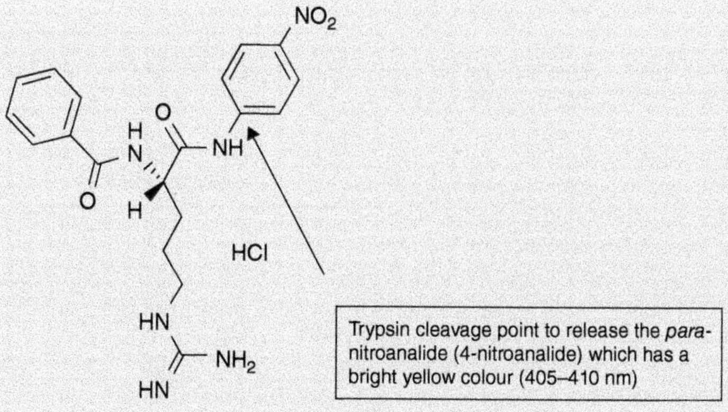

Figure 3.12 Benzoyl L-arginine 4-nitroanalide (also known as benzoyl L-arginine para nitroanalide; BAPNA).

We do not have a 1000 ml reaction volume. Therefore, adjusting for 1.0 ml by dividing by 1000

$$= 20.3 \, nmol \, ml^{-1} \, min^{-1}$$

We do not have a 1.0 ml volume; we have a 0.2 ml volume in the microplate well

$$= 20.3 \, / \, 1000 \text{ to arrive at } 1 \, \mu l \times 200 \text{ to arrive at } 0.2 \text{ ml volume in the microplate well}$$

$$= 4.06 \, nmol \text{ in the well } min^{-1}$$

International units of enzyme activity is 1.0 μmol of product formed min⁻¹

Therefore, we have
a. 0.00406 international units of enzyme activity in the microplate well.
   As these units have been derived from 10 μl of enzyme, therefore we have

$$0.00406 \text{ units} \times 100 \text{ to give us units } ml^{-1} \text{ of enzyme} = 0.406 \text{ units } ml^{-1} \text{ of enzyme}$$

b. Total protein $= 25 \, ml \times 0.05 \text{ mg } ml^{-1}$
   $$= 1.25 \text{ mg of protein}$$

c. Specific activity $= \dfrac{\text{units } ml^{-1}}{\text{protein mg } ml^{-1}}$

   $$= \dfrac{0.406}{1.25}$$

   $$= 0.325 \text{ units mg}^{-1} \text{ protein}$$

d. Total activity $= \text{units } ml^{-1} \times \text{total volume } (ml)$
   $$= 0.406 \times 25$$
   $$= 10.15 \text{ units in total}$$

Worked Example 3.6 The Assay of a Peptidase (Trypsin) Using Azocasein as the Substrate Measured Using a 1.0 ml Cuvette

Trypsin cleaves peptide bonds when the peptide or protein has a lysine or arginine residue. The cleavage leaves two smaller peptides one with a lysine or arginine amino acid at the carboxy (–COOH) end of the peptide. The other peptide with a free amino end (–NH₂) will have several different amino acids, depending on where in the primary sequence the cleavage has occurred.

As the presence of the basic amino acids lysine and arginine in a protein is ubiquitous, trypsin can cleave all proteins. Therefore, the trypsin assay with azocasein closely mirrors the action of trypsin in the lower intestine on whole or partially digested proteins.

When using proteins as substrates in an enzyme assay, the molar absorptivity coefficient (ε) used for transparent coloured solutions is replaced with a mass absorptivity coefficient ε 1%(w/v) at a specific wavelength.

A 1% (w/v) solution of a protein is 1.0 g dissolved in a 100 ml of water or buffer. This concentrated solution of protein will absorb light at a given wavelength. Azocasein in an alkali environment absorbs light at 440 nm.

The mass absorptivity coefficient for azocasein is $\varepsilon_{440 \, nm}^{1\%(w/v)} = 35$

This value means that a 1% (w/v) solution of azocasein will give an absorbance at 440 nm of 35.

We know that most spectrophotometers have a limit to their detection between 0 and 2.0.

Therefore, the absorbance of a 1% (w/v) azocasein solution is above the detection limit of the spectrophotometer, i.e. greater than an absorbance of 2.0 (see Section 3.7.5).

A 0.1% (w/v) solution of azocasein at 440 nm solution will give an absorbance of 3.5, which is still too high for the spectrophotometer to detect.

A 0.01% (w/v) solution of azocasein at 440 nm will give an absorbance of 0.35; this is within the detection range (0–2.0) of the spectrophotometer.

Therefore, the range of azocasein concentrations that can be detected using a spectrophotometer is approximately 0.05–0.002% (w/v) with an absorbance of 1.75–0.17 at 440 nm.

Example: To assay the cleavage of internal peptide bonds by trypsin a solution of 2% (w/v) azocasein protein solution is prepared and stored at 4 °C until required.

Assay mixture:

0.25 ml of 2% (w/v) azocasein (orange) is added to a microfuge tube along with 0.25 ml of 100 mM Tris/HCl (pH 8.3) containing 5 mM $CaCl_2$, 0.15 ml of water and 0.25 ml of 0.05%(w/v) trypsin.

The 0.9 ml mixture was incubated on a heating block for 30 min at 37 °C. The reaction was terminated by the addition of 0.1 ml of 72% (w/v) TCA to give a final TCA concentration of 7.2% (w/v). This acidic environment will precipitate all the proteins (azocasein and trypsin), but it will leave the smaller cleaved peptides ($Mr < 3000$) with the coloured azo dye attached in solution as a measure of how many peptides have been released from azocasein by trypsin (peptidase)

The reaction tubes are then placed on ice for 10 minutes to encourage protein precipitation.

The precipitated proteins can be removed from the coloured peptides by centrifugation at $15\,000 \times g$ (RCF) for 15 minutes.

To determine the peptidase activity, the tubes with trypsin present are compared to a solution which has been precipitated at the start of the experiment (time zero).

The control values of the reaction mixture precipitated at time zero by the addition of 72% (w/v) can be subtracted from the absorbance values obtained after trypsin digestion of azocasein for 30 min at 37 °C.

To develop a vivid orange colour 0.5 ml of the supernatant remaining after centrifugation is added to 0.5 ml of 4.0 M NaOH in a 1.0 ml cuvette to measure the orange colour at 440 nm.

Alternatively, fill the wells of a microplate with 0.1 ml of 4.0 M NaOH and then add 0.1 ml of the supernatants to the plate in triplicate. The absorbance is then read at 440 nm.

For example: The time zero values at 440 nm gave values of 0.05. The enzyme treated samples gave values of 0.56, 0.52 and 0.54.

The average value of the enzyme catalyzed samples was 0.54

The value minus the zero-time control = 0.49

The change in absorbance at 440 nm over 30 minutes = 0.49 / 30

$$= 0.016\,min^{-1}$$

Using the Beer–Lambert law and a 1.0 ml cuvette

Change in absorbance min^{-1} = the pathlength of the cell (cm) × mass absorptivity coefficient $(\varepsilon^{1\% \,(w/v)})$ × percentage concentration (g in a 100 ml).

$$0.016 = 1 \times 35 \times \text{concentration}\left(\text{g in a 100 ml}\right)$$

$$\frac{0.016}{35} = \text{concentration}\left(\text{g in a 100 ml}\right)$$

= the concentration of peptide generated is 0.00046 g of azocasein peptides in 100 ml

Therefore, in the 1.0 ml cuvette used the amount of azocasein peptides = 0.0000046 g or 4.6 µg of azocasein peptides generated min^{-1} by the action of trypsin.

The international unit of enzyme activity is 1 µmol of product formed min^{-1}. However, for assays that use a protein as a substrate the international units can be defined as 1 µg of product formed min^{-1} at a specified wavelength.

In this assay, we have 4.6 µg of azopeptides generated or 4.6 units of enzyme activity from 0.5 ml of the enzyme-digested extract.

Therefore, in 1.0 ml extract, we would have 9.2 units of peptidase activity.

All this product has been derived from the 0.25 ml of enzyme that was used in the original experiment.

Therefore, in our original stock of 0.05% (w/v) trypsin, we have 9.2 × 4 = 36.8 units ml^{-1} of enzyme activity using azocasein as the substrate.

3.9 Summary

- Light in the UV/visible light (200–800 nm) part of the electromagnetic spectrum can be measured in a spectrophotometer.
- The absorbance of light is related to the concentration of a solution and the path length the light must travel through the solution.
- The biological molecules in solution can be detected using a spectrophotometer and quantified using the Beer–Lambert Law.

Notes

1   For example, compounds that fluoresce have a wavelength number at which they maximally absorb photons (excitation wavelength) and a wavelength at which they re-emit a photon (emission wavelength). This is described in Section 3.3.1.

2   Worked example 3.1 demonstrates that longer wavelengths of the electromagnetic spectrum have less energy.

3   Molar concentrations are not always suitable for proteins and 1% (w/v) mass absorptivity at a specified wavelength at 25 °C is quoted (for example, bovine serum albumin (BSA) at 280 nm $\varepsilon^{1\%\,(w/v)} = 6.6$). The molar absorptivity coefficient for bovine serum albumin at 280 nm is 43 824 M^{-1} cm^{-1}.

4   The surface of a CD or DVD is etched with many closely spaced reflective lines. These discs can act as a diffraction grating; if they are angled in a beam of white light, the colours of the rainbow can be seen.

5   A chromophore is the part of the molecule that is responsible for the colour which can be seen or measured in a spectrophotometer.

4

DATA ANALYSIS AND PRESENTATION

4.1 Introduction

A good laboratory report is one that contains data that have been analyzed appropriately and presented effectively. This chapter gives you an introduction to some of the key elements of basic statistical analysis applied to data in the biosciences and provides hints and tips on how to present your data effectively in laboratory reports.

Although you may do some experiments in laboratory classes only once, it is important to realize that you would be unlikely to get exactly the same result if you repeated the experiment. This is because, with biological measurements, there is always variation or spread of the data. For this reason, it is important to establish the reliability of data in terms of the accuracy and precision of the values obtained from bioscience experiments.

Definitions: Accuracy and Precision

Accuracy
- This refers to the proximity of a measured value (or an estimate from a collection of measurements) to the actual value.
- In the laboratory, this can be determined by comparing the experimental output from the test or assay in question with those of a reference method.

Precision
- This refers to the repeatability of the experimental measurements, as measured by the spread of data which reflects the reproducibility of measurements. When experiments are repeated several times independently, narrow error bars are an indication of good reproducibility.
- Ideally, an experiment should produce readings that are both accurate and precise.

Worked Example 4.1 Activity Measurements

If the specific activity of an enzyme measured by a long-established method is 12 units per mg protein, which would be the most suitable alternative based on the measurements from three potential new assay methods (A, B and C) giving the specific activities indicated below:

A. 12.4 ± 0.6
B. 12.1 ± 0.1
C. 12.0 ± 0.6

- Method A is unsuitable as it gives values that are inaccurate and imprecise.
- Although the mean value from method C is the most accurate, it is not the most precise as indicated by the wide spread of data compared to method B.
- Method B may be the most suitable alternative from the three methods because it gives values within 1% of the expected values (accurate) with a very narrow spread of data (precise).

Basic Bioscience Laboratory Techniques: A Pocket Guide, Second Edition. Philip L.R. Bonner and Alan J. Hargreaves.
© 2022 John Wiley & Sons Ltd. Published 2022 by John Wiley & Sons Ltd.

In a significant piece of research, such as a research project, effective assessment of data requires statistical analysis, which could take various forms including basic descriptive statistics of data (e.g. estimation of midpoint and spread of data), to assessment of the relationships between different sets of data or variables (e.g. time, concentration, temperature, etc.). However, for effective communication of your findings, it is also important to present your data clearly and unambiguously in your laboratory or project report.

Properly used, statistical analysis provides valuable support for the scientific validity of the findings. However, improperly applied statistical analysis can be misleading, as indicated by the numerous quotes you will see at discrete points in this chapter.

Statistics

'There are three kinds of lies: lies, damn lies and statistics' – Attributed to Benjamin Disraeli by Mark Twain.

'Facts are stubborn, but statistics are more pliable' – Mark Twain.

'Definition of statistics: The science of producing unreliable facts from reliable figures' – Evan Esar.

- These quotes refer to the common skepticism at the use of statistics to justify political decisions or to criticize those of opponents.
- This reflects the selective use of statistics to support a specific argument rather than the appropriate use of statistics to analyze the validity of data.
- The moral of the story is that an appropriate statistical analysis should be applied to the data in question and the method used should be declared openly and honestly in any publication or presentation using such data so that it can be open to scrutiny.
- In most areas of science, there is not always agreement on the methods used to analyze data.

'Definition of a statistician: A man who believes figures don't lie but admits that under analysis some of them won't stand up either' – Evan Esar.

4.2 Statistical Analysis of Data: Some Key Definitions

4.2.1 Populations and Samples

In statistics, the term 'population' refers to very large sets of data that we would ideally want to analyze (i.e. large numbers of measurements). However, it is simply not practical to analyze large populations. Instead, a *sample* is studied, which contains a smaller number of data points 'picked' from the population, which is assumed to be representative (i.e. an estimate) of the whole population.

4.2.2 Variables and Observations

A *variable* is a characteristic that differs from one member of the population and the next and they can take various forms such as qualitative (e.g. observed or not observed) or quantitative (e.g. specific activities of an enzyme, protein concentration, etc.).

Different types of statistical analysis are required to analyze samples containing these distinct types of data. As a minimum requirement, some form of descriptive statistics should be performed on your data.

4.2.3 Descriptive Statistics

This term refers to a number of analyses that can be applied to describe the midpoint and spread of data within a sample (see Worked Example 4.2). Estimates of midpoint include:

Median - This is the value in the middle when numbers have been listed in rank order.
Mode - This is the most frequently occurring observation.

Mean - The mean is the arithmetic average of all values in the sample or population. To calculate the mean, we add all values together and divide the sum by the number of values included.

Worked Example 4.2 Estimation of Midpoint Values

Median
- Calculate the median of the set of values 1, 2, 3, 4, 5.
- To do this, look for the obvious midpoint in the values, if one exists. In this case, if the values are placed in increasing or decreasing order, the clear midpoint value is 3.
- If the values were 2, 4, 8, and 10, the median would be 6 (in this case, the two middle numbers in the group are averaged).
- Thus, in the event of an even number of values in the sample, the median would be the average of the middle two values.

Mode
- Estimate the mode in the following set of values: 1, 5, 3, 5, 7, 5, 5 and 8.
- To do this, simply measure the frequency of each value and the one that appears most frequently is the mode.
- In this case, the value 5 appears four times and is clearly the mode, as the other values only appear once.

Mean
- Estimate the sample mean for the following data: 23, 13, 20, 32, 5, 10, 7, 25, 11 and 18.
- To do this, calculate the sum of the ten values, then divide by the number of values (10).
- = 164/10 = 16.4.
- As discussed earlier, this value would effectively be our estimate of the population mean, which would apply to a much larger number of data values.

Student exercises

- Calculate the median for the following set of data: 3, 8, 10, 12.
- Calculate the mode for the following set of data: 5, 6, 7, 8, 9, 7, 10, 11, 7, 12, 13, 12.
- Calculate the mean of the following set of data: 21, 15, 18, 17, 25.

4.2.4   Measures of Spread (Dispersion)

Different sample populations may have the same mean, median or mode but differ in the *spread* of data. Variability is likely to be greater in small sample sizes and, as the mean of the sample is an estimate of the population mean, an estimate of the variability of the sample data can give us an idea of the accuracy and precision of the calculated mean value.

There are various measurements of variation of relevance to bioscience research data, which are indicated below. These include:

Range: Highest and lowest values, and so on.

Variance: (S^2) can be determined from the following equation:

$$S^2 = \frac{1}{n-1} \sum_{i=1}^{n} (x - \overline{x})^2$$

where n represents the number of measurements, i is the ith value, x is each individual value and $\overline{x}$ is the mean of all measurements. This is an estimate of the *population* variance, which is normally referred to as σ^2.

Standard deviation (s): This is the square root of the variance $\left(\sqrt{s^2} \right)$.

One formula to estimate the sample standard deviation (s) is:

$$s = \sqrt{\frac{1}{n-1}\sum_{i=1}^{n}(x-\bar{x})^2}$$

The sample standard deviation is an *estimate* of the population standard deviation $\left(\sqrt{\sigma^2}\right)$. For each data point, deviation from the mean is calculated as the difference between that value and the mean, which could be a positive or negative value. Typically, standard deviation values are represented as error bars for (mean) values of data presented in graphs or histograms. Alternatively, data may be presented in tables as mean ± SD.

On the other hand, *for small sample sizes*, the standard error of the mean (SEM) is often used to estimate the spread of data. This is derived by dividing the standard deviation by the square root of the number of values (n), as shown below:

$$\text{SEM} = \frac{s}{\sqrt{n}}$$

Such values give an indication of the spread of a population's data about its mean. As a rule, the spread of data decreases as the population size increases. Thus, small sample sizes may result in a broader distribution curve, reflecting wider standard deviations.

You should try to calculate these values using the formulae above so that you can understand better the principles involved (see Worked Example 4.3). However, there is a variety of statistical packages for PCs that can be used to perform basic descriptive statistics (e.g. Microsoft Excel, GraphPad Prism, SPSS, Minitab, etc.).

Worked Example 4.3 Calculation of Mean and Standard Deviations

Find the mean and standard deviation from the following set of values (x): 23,13,20,32,5,10,7,25,11,18.

To do this, construct a table with the headings x, $x-\bar{x}$ and $(x-\bar{x})^2$, and fill in the individual values given above, their differences from the mean (which may be positive or negative values) and the square of the mean differences (always positive).

- These values can then be substituted into the equations above to calculate the standard deviation.

X	$x - \bar{x}$	$(x-\bar{x})^2$
23	6.6	42.56
13	−3.4	11.56
20	3.6	12.96
32	15.4	237.16
5	−11.4	129.96
10	−6.4	40.96
7	−9.4	88.36
25	8.6	73.96
11	−5.4	29.16
18	1.6	2.56
	TOTAL	669.20

- Calculate the mean by summing all of the values and dividing by the number of values.

$$\text{Mean}(\bar{x}) = \frac{(23+13+20+32+5+10+7+25+11+18)}{10} = \mathbf{16.4}$$

- Now, calculate the standard deviation, incorporating the calculated value from the sum of the squares of the differences from the mean (i.e. 669.2)

$$s = \sqrt{\frac{1}{(10-1)} \times 669.2} = \sqrt{74.36} = 8.623$$

Standard deviation = **8.623**

- The standard error is then obtained by substituting the standard deviation value in the following equation:
- SEM = SD/√n = 8.623/3.162 = **2.727**

Student exercise

Calculate the SD and SEM values for the following data set: 46, 26, 40, 64, 10, 20, 14, 50, 22, 36.

4.3 Distributions

While basic descriptive statistics will give us an idea of the midpoint and spread of data points within a sample, in order to establish the significance of any differences observed between basic data from two or more different samples, it is necessary to apply an appropriate statistical test. The type of test that should be applied is, in turn, dependent on the nature of the distribution of values within the samples. Several types of distribution exist as discussed in the following sections.

4.3.1 Normal Distribution

In a population with a normal distribution, the individual values are distributed symmetrically about the mean, giving a bell-shaped curve (see Figure 4.1a). In theory, a perfect normal distribution is only likely to occur in large samples and populations, although it is possible to determine the likelihood that a smaller sample approximates to a normal distribution. In practice, the normal distribution is converted to a standard normal distribution (or *z distribution*) by introducing a correction factor to give each sample population a mean of zero, using the formula:

$$z = \frac{X - \mu}{\sigma}$$

where X is an individual score, μ is the mean and σ is the standard deviation of the original normal distribution. In the z distribution, 96% of values lie within the mean ± 2 SD units and 68% within the mean ± 1 SD unit (see Figure 4.1b). Therefore, calculation of the z value (see Worked Example 4.4) for a particular measurement will give an indication of by how many standard deviation units it differs from the population mean. If the value is more than 2 SD units above or below the mean, it falls outside the 96% range.

Worked Example 4.4 Z-value Calculation

Find the Z value for the mark of a student scoring 35 in a test for which the average mark was 25 and the standard deviation was 5.

$$Z = (35 - 25)/5 = 2$$

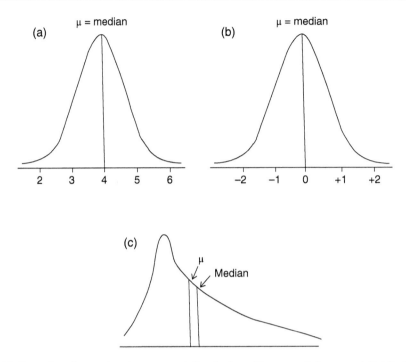

Figure 4.1 *Examples of normal and non-normal distributions. Shown are (a) a normal distribution, (b) a standard normal (or z) distribution and (c) a non-normal distribution.*

- This means that the mark obtained was the equivalent of two standard deviation units higher than the class average.
- This value together with z values for the other students' marks would be used to construct the standard normal distribution of the class marks.

Student exercise

A student achieved a mark of 73% in a class test for which the average mark was 50% and the standard deviation was 11.5. Find the z value and indicate the number of SD units that the student's mark differs from the class mean.

A randomly drawn element from the population has a 4% probability of falling within the outer range. As it is typical to apply confidence limits of 95%, the 95% confidence interval can also be expressed, which represents the range within which 95% of values will be located if the mean of the sample is not significantly different from the population mean (μ). In general, the narrower the calculated range, the higher the level of confidence in the accuracy of the estimation.

The confidence interval (CI) is calculated as follows:

$$\text{CI} = \text{mean } plus \text{ } or \text{ } minus \text{ t value} \times \text{standard error}$$

Confidence interval calculations can form an important part of more complex statistical tests such as analysis of variance (ANOVA), which can be used more effectively for the comparison of multiple sets of data, which will be discussed later. However, an example of confidence interval estimation to determine whether sample means are likely to be different from each other is shown below. If the confidence intervals do not overlap, the means are likely to be significantly different from each other.

This example contains data sets from an experiment in which the effects of a toxin were studied on the outgrowth of axons by cultured nerve cells at two distinct time points (4 and 24 hours).

It should be noted that, in the worked example, the wider spread of values in toxin-treated cells means that a more robust statistical test should be applied, as discussed later.

Worked Example 4.5 Confidence Intervals

This example contains data sets from an experiment in which the effects of a toxin were studied on the outgrowth of axons by cultured nerve cells at two distinct time points (4 and 24 hours).
Determine the confidence intervals for the data below.

- Statistical packages may provide a plot of these: If they do not overlap, the means are likely to be different.
- However, if you check the probability level, they may be separated at 0.05, but not at 0.01, giving an idea of how significantly different the data sets are from one another.

Confidence intervals are the mean $\pm$ t $\times$ standard error, where the value of t is looked up in a statistical table for $n - 1$ degrees of freedom and the desired level of probability. For example, in the case of three replicates, $n - 1 = 2$; for a probability of 0.05 (which sets confidence intervals to 95%), t = 4.3 (see Section 4.4.1).
For each column of data, the standard error of the mean or SEM = $\sqrt{s^2/n}$

Statistic	Control 4 h	Toxin-Treated 4 h	Control 24 h	Toxin-Treated 24 h
variance	19.59	7.47	16.00	5.13
SEM	2.21	1.37	2.31	1.13
t × SE	9.50	5.89	9.89	4.86
mean	32.35	13.23	35.90	18.50
Confidence interval at 95% limits	22.85 → 41.85	7.34 → 19.12	26.01 → 45.79	13.64 → 23.36

- Initial observation of these confidence limits in the bottom row of the table indicates that there is no overlap between treatments and control at a given time point.
- Therefore, it is likely that the toxin is having an inhibitory effect on axon growth.
- These limits can be plotted manually or on statistical packages (see below for a typical plot in Minitab), to have a more visual confirmation of the lack of overlap.
- However, the overlap between controls at 4 and 24 hours suggests that there is no significant increase in axon numbers in controls from 4 to 24 hour incubation.

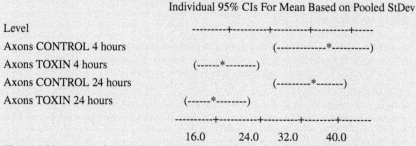

Individual 95% CIs For Mean Based on Pooled StDev

```
Level                        ---------+---------+---------+---------+-----
Axons CONTROL 4 hours                       (-------------*----------)
Axons TOXIN 4 hours          (------*--------)
Axons CONTROL 24 hours                      (---------*-------)
Axons TOXIN 24 hours         (------*--------)
                             ---------+---------+---------+---------+--------
                                16.0      24.0     32.0      40.0
```

- The variable spread of values between treatments means that a more robust statistical test should be applied, as discussed later.

4.3.2 Asymmetric (Non-Normal) Distribution

In this type of distribution, the sample values are distributed in a more skewed fashion (see Figure 4.1c), which is more typical in small sample sizes. For statistical analysis, samples showing such distributions should be compared with each other about their midpoint, in this case the median. Statistical analysis to determine the similarity between two samples with non-normal distributions would normally involve the use of a non-parametric statistical test such the Mann Whitney U test.

4.3.3 Discrete Probability Distributions

The binomial distribution is another common type of distribution that you may come across in your early studies at university. This applies to a situation where there are two possible equally likely outcomes that are completely independent of each other (e.g. success/failure, survival/death, male/female, etc.) and have an equally likely chance of occurring if trials (i.e. the process of selecting an individual at random from the population) are done randomly, independently of each other and in sufficient numbers.

The Poisson distribution is important in the analysis of certain types of data associated with randomness. It is used in situations where nothing can influence the data except chance. In this case, the Poisson distribution would describe the number of counts we would expect per unit time or space and the data would present a Poisson distribution around the mean of the fitted values. The Poisson distribution is also discrete in that it is suitable for describing count data.

For example, the likelihood of obtaining a head after tossing a coin is, in theory, 0.5 (50%). In practice, this means that if the coin is tossed many times, the outcome should be 50% heads and 50% tails ($p = 0.5$). Similarly, in the case of rolling a dice, the probability of obtaining a specific value (e.g. 1) is 1 in 6, whereas the probability of obtaining any number but one on a given trial would be 5/6.

Probability

'If a coin falls heads repeatedly one hundred times, then the ignorant would claim that the "law of averages" must almost compel it to fall tails next time, any statistician would point out the independence of each trail and the uncertainty of each outcome. But any fool can see that the coin must be double-headed'. – Ludwick Drazwk.

'The statistics on sanity are that one out of every four Americans is suffering from some kind of mental illness. Think of your three best friends. If they're OK, then it's you'. – Rita Mae Brown.

'The buffalo isn't as dangerous as everyone makes him out to be. Statistics prove that in the United States more Americans are killed in automobile accidents than are killed by buffalo'. – Art Buchwald.

'Statistics show that of those who contract the habit of eating, very few survive' – William Wallace Irwin.

So, as you can see, you shouldn't believe everything you read about statistics!

4.4 Statistical Comparison of Data

In order to determine whether sets of data are statistically different from each other, it is convention to apply the null hypothesis (H_0), which usually states that there is no difference between the data samples (i.e. they are likely to have arisen from the same population distribution – H_0: $\mu = 0$). The alternative hypothesis is that there is a difference, although the direction of that difference is rarely stated (i.e. H_0: $\mu \neq 0$). In principle, whatever test is applied (which depends on the precise nature of the data and its distribution), the null hypothesis is not rejected unless there is convincing evidence against it. Both the null and the alternative hypotheses are expressed in a manner that reflects hypothetical population parameters (e.g. as it is unlikely that the sample mean will be equal to zero). One tailed and two tailed tests can also be applied, depending on whether any differences from the mean are expected to occur in a specific direction (i.e. an increase *or* decrease) or whether they could occur equally in either direction, respectively.

It is typical to use 95% confidence limits when performing a statistical test in bioscience research. In other words, the calculated probability that the two samples belong to the same population must fall below 1 in 20 or 5% (normally expressed as $p < 0.05$) in order that the null hypothesis can be rejected. The lower the probability value falls below the 0.05 level, the more highly significant the difference. For example, a value of $p < 0.01$ would indicate that there is less than a 1% chance that the samples being compared could have arisen from the same population, and so on. On the other hand, a value above 0.05 would be taken as evidence that the two sample distributions could have resulted from chance variation within the population are not therefore likely to be significantly different from each other. However, if sample measurements in biomedical studies show significant levels of variability and the p value is approaching significance (for example, 0.05–0.1), significance may be eventually achieved by increasing the number of measurements in each sample (particularly if the sample sizes are on the small side). Let us now look at the basis of some of the more common statistical tests used on bioscience data, and how they are applied to different sample distributions and data types.

4.4.1 Comparing Two Sets of Values

4.4.1.1 Unpaired and Paired Parametric Tests (t Tests)

This type of test (see Worked Example 4.6) is applicable to samples showing a normal distribution and is often applied in biomedical research studies. The unpaired t test is typically used to compare two independent samples about their arithmetic means. Prior to using this test, it is advisable to check that both sets of data are normally distributed, as this is not always the case for small sample sizes. If the test is applied to samples with non-normal distributions, the confidence limits at which rejection of the null hypothesis is set may not be reliable. Thus, conclusions drawn from data showing significant or non-significant changes with p values close to the chosen confidence limits may be wrong if the wrong significance test is applied. Having said that, the test may still work well when distributions deviate from normality, provided that they are reasonably symmetrical and continuous. Therefore, very low p values (e.g. 0.001) are still likely to reliably reflect that there exists a significant difference between two data sets.

The test is based on the t distribution of data. The t value for each data point can be obtained from the equation:

$$t = \frac{\bar{x} - \mu}{s / \sqrt{n}}$$

This is an approximation of the z distribution described earlier (see Chapter 4, Section 4.3.1), taking into account the sample size. The shape of the t-distribution approximates to the z distribution for large sample sizes, whereas the curve becomes flatter at lower sample number (n). The calculated t value can be read off in statistical tables of t-distributions, which take into account the sample size and distribution by correcting for the 'degrees of freedom' for a particular set of data, which is determined by $n - 1$ (where n is the number of measurements). This factor is normally taken into account when using a computer package (e.g. GraphPad Prism or Minitab) to perform statistical tests. Otherwise, you can look up the value in statistical tables, which can be found online or purchased (see further reading section). With this test, you are effectively estimating the probability that the two samples could have arisen by chance from the same population distribution. A value of $p < 0.05$ would be indicative of a significant difference between the two sets of data.

Worked Example 4.6 Student's t Test

- Determine whether the following randomly taken set of measurements (12, 24, 13, 5, 10, 18, 8, 45, 14 and 13) could be part of a normal distribution with a mean of 6.
- First of all, estimate the *sample mean* as follows:

$$(12 + 24 + 13 + 5 + 10 + 18 + 8 + 45 + 14 + 13)/10 = 162/10 = \mathbf{16.2}$$

If we then estimate the *standard deviation*, as described in the earlier worked example:

X	$x - \bar{x}$	$(x - \bar{x})^2$
12	−4.2	17.64
24	7.8	51.84
13	−3.2	10.24
5	−11.2	125.44
10	−6.2	38.44
18	1.8	3.24
8	−8.2	67.24
45	28.8	829.44
14	−2.2	4.84
13	−3.2	10.24
	TOTAL	**1158.6**
	Square of mean differences	

The standard deviation of the sample is calculated as follows:

$$s = \sqrt{\frac{1}{(10-1)} \times 1158.6} = \sqrt{128.73} = 11.35$$

The t value is then calculated as follows:

$$t = \frac{\bar{x} - \mu}{s/\sqrt{n}}$$
$$= (16.2 - 6) \div (11.35 \div \sqrt{10})$$
$$= 10.2 \div 3.639$$
$$= 2.802$$

- If we look at a table of t distributions, we can see that the 5% point for t is 2.26 at 9 degrees of freedom $(n-1)$, which means that 95% of the distribution will lie with a t value of −2.26 to + 2.26.
- As the observed t value falls outside the central 95% of the distribution, we can *reject* the null hypothesis at the 5% level, as we cannot reasonably assume that the sample was drawn from a population with a true mean of 6.0.

In order to compare two *sample* means, a similar approach is adopted taking into account the central limit theorem. Thus, if two independent biological samples (with means $\bar{x}$ and $\bar{y}$, containing N_1 and N_2, values respectively) are taken from a population with a normal distribution, they too are assumed to be normally distributed with a variance of σ^2/N_1 and σ^2/N_2.

The difference $(\bar{x} - \bar{y})$ contains two varying elements (i.e. the two-sample means) both of which contribute to its total variance, such that the variance for $(\bar{x} - \bar{y})$ is equal to $(\sigma^2/N_1) + (\sigma^2/N_2)$.

We can then propose the null hypothesis that both $\bar{x}$ and $\bar{y}$ come from the same parent population (with mean and variance of σ^2 and μ), in other words that they are not significantly different from each other. In order to do this, the pooled variance must first of all be calculated from the equation:

$$S^2 = \frac{1}{N_1 + N_2 - 2} \times \left[\left(\Sigma x - \bar{x} \right)^2 + \left(\Sigma y - \bar{y} \right)^2 \right]$$

Then, calculate the t value for $(N_1 + N_2 - 2)$ degrees of freedom from:

$$t = \frac{\bar{x} - \bar{y}}{\sqrt{s^2 \left(\dfrac{1}{N_1} + \dfrac{1}{N_2} \right)}}$$

The t value as before can be read off (as explained earlier) in statistical tables for the appropriate number of degrees of freedom to determine whether the two sample means lie within the same distribution.

Alternatively, if the two sets of data to be compared are *dependent* on each other, a *paired t test* can be performed. In this case, the null hypothesis is that the mean of the differences between the paired sets of observations is zero (i.e. that the distribution of mean differences is not significantly different from that of a population with a mean of zero). In biomedical research, an example of paired data could be the use of one set of patients to test the effects of two different drugs on a specific parameter such as blood pressure.

$$t = \frac{\bar{d} - \mu d}{\overline{sd}}$$

In this equation, $\bar{d}$ is the mean of the differences between paired observations, μd is the hypothesized mean difference between paired observations (usually zero) and $\overline{sd}$ is the standard error of $\bar{d}$. As for unpaired t tests, in the paired sample t test, the value for degrees of freedom is $n-1$, with n representing the number of *pairs* of observations. A one sample t test is applied, as we have effectively reduced the data to one variable by dealing with the difference between each paired observation. For the test to be a valid indicator of significance, it is assumed that the distribution of the differences between individual pairs is normal and that the data were randomly selected from the distribution.

4.4.2 Unpaired and Paired Non-Parametric Tests

Most of your current needs for statistical testing will probably be covered by one or a combination of the tests described so far. However, the Mann Whitney U test is often applied to unpaired data sets that exhibit deviation from a normal distribution, which is often the case with small sample sizes. It compares the samples' distributions about the median rather than the arithmetic mean, which considers the asymmetric spread of the data. Although paired t tests are often used irrespective of the mean difference distribution, if the data sets are paired but exhibit a non-normal distribution, another non-parametric test should be applied, such as the Wilcoxon paired sample test. This test checks the number and sizes of positive and negative differences between paired data by putting them in rank order, summing the ranks of the positive and negative differences. The rank sums provide the test statistic which is then subjected to a binomial test to determine the p value. The null hypothesis is that there is no difference between the samples; therefore, the sum of the positive and negative differences should be zero. This test can be applied using the One-sample Wilcoxon test, which is available in most statistical software packages.

4.4.3 Comparisons of Multiple Sets of Data

It is, of course, not always the case that only two data sets are being compared with each other. In fact, in bioscience research, it is quite common to do a statistical comparison between three or more groups of data. This could, for example, be an experiment in which several drug or toxin concentrations are compared against the same control treatment. In some cases, there may be two or more variables (e.g. incubation time, temperature and concentration). When comparing multiple sets of data, it is not considered reliable to perform multiple tests comparing different pairs of data sets, as there is a greater rate of both type I and type II errors, in which an incorrect inference of statistically significant change or no significant change is obtained, respectively.

One of the most applied multiple comparison tests is ANOVA. If the null hypothesis is rejected after ANOVA (i.e. $P < 0.05$), we know that there is one or more statistically significant difference between the groups, but we do not know exactly where those differences are. It is customary to perform a further '*post-hoc*' analysis to pinpoint significant differences between specific experimental groups.

There are several common multiple comparison tests, including the Tukey method, the Newman–Keuls method, the Bonferroni method and the Dunnet method. These methods are typically used in statistical analysis software packages such as Minitab and GraphPad Prism. When using one of these packages for ANOVA, remember to select the appropriate multiple comparisons post-ANOVA analysis to confirm exactly which groups show significant differences from each other.

If a single variable (e.g. concentration range) is compared to the same control, one-way ANOVA can be applied. If there are two or more experimental variables, two-way ANOVA or multiple ANOVA (MANOVA) must be applied, respectively. Be especially careful to follow an available online tutorial on how to enter your data for such a statistical test.

The ANOVA test makes a number of assumptions that all of the experimental data groups should meet for the test to be appropriate, including equal variances and a normal distribution in all data sets. If these assumptions are not met, a non-parametric multiple comparisons test, such as the Kruskall–Wallace test, should be applied.

4.4.4 Chi Square Test for Dichotomous Variables

Binomial distributions are distinct two-class populations (e.g. success/failure; dead/alive; sick/well; male/female). In this case, comparison can be made using the chi square test, which compares observed with expected outcomes. In a simple chi square test, the data are set out in tabular form:

Group	Success	Failure	Total
A (observed)	a	b	a + b
B (expected)	c	d	c + d
TOTAL	a + c	b + d	n

$$\text{Chi Squared}\left(\chi^2\right) = \frac{\left(ad - bc\right)^2 \times n}{\left(a + c\right)\left(b + d\right)\left(a + b\right)\left(c + d\right)}$$

In this particular case, the final chi squared value must be 3.84 or above to indicate a significant difference at $p < 0.05$ and 6.64 or above for $p < 0.01$. These values can be found in statistical tables and vary according to the number of degrees of freedom, which is in turn related to the number of measurements. As the table above contains two measurements (success or failure), the number of degrees of freedom = 1 (i.e. $n - 1$). This is taken into account when reading chi square values in statistical tables, although a computer package would do this automatically.

Worked Example 4.7 Chi Square Test

A coin was tossed 100 times, with the following results (observed counts). Use the chi square test to determine whether these counts differ significantly from the expected counts (i.e. equal numbers of heads and tails). The null hypothesis would be that there is no significant difference between the two sets of counts (i.e. the coin is not biased towards heads or tails).

	Heads	Tails	Total
Observed counts (A)	54 (a)	46 (b)	100 (a + b)
Expected counts (B)	50 (c)	50 (d)	100 (c + d)
TOTAL	104 (a + c)	96 (b + d)	200 (n)

$$\chi^2 = \frac{(ad-bc)^2 \times n}{(a+c)(b+d)(a+b)(c+d)}$$

$$\chi^2 = \frac{(ad-bc)^2 \times n}{(a+c)(b+d)(a+b)(c+d)}$$

$$\chi^2 = \frac{(2700-2300)^2 \times 200}{104 \times 96 \times 100 \times 100}$$

$$\chi^2 = 32\ 000\ 000\ /\ 99\ 840\ 000 = \mathbf{0.321}$$

Degrees of freedom $= 1$

In order to represent a significant difference from the expected value at $p \leq 0.05$ or $p \leq 0.01$, the chi square values would need to be ≥ 3.84 or 6.64, respectively, as can be seen from statistical tables.

Therefore, in this case, there is no significant difference from the outcomes that would be expected by chance.

4.4.5 Analyzing Relationships Between Two Variables: Linear Correlation and Regression

The methods you are most likely to use to study the relationship between two variables are linear regression and linear correlation, both of which are parametric tests with certain requirements that must be met. Regression is used to describe how an independent variable (X) can cause changes in a dependent variable (Y), allowing us to predict the value of Y if we know the value of X. This is particularly useful when reading off Y values on a calibration graph (e.g. as part of a protein estimation or an enzyme assay). However, correlation simply helps to describe two variables that are associated with each other and vary together. It does not necessarily prove a cause and effect relationship between the two variables.

Simple linear regression helps to describe a causal or predictive relationship between two variables using a straight-line graphical plot. The equation of the straight line is as follows:

$$\mathbf{y} = \alpha + \beta \mathbf{x}$$

In this equation, 'x' represents the independently measured variable (*explanatory* or *predictor* variable), whereas 'y' is the dependent variable (i.e. the *response* variable), 'α' is the intercept on the y axis (i.e. the mean value of y when x = 0) and 'β' is the gradient of the line (i.e. the changes in y values per unit change in x).

However, for linear regression to be applicable, certain conditions have to be met:

Firstly, a straight line must be the true relationship between the two variables. Secondly, for any value of x, there must be a distribution of possible y values, the mean of each of which falls on the straight line, and the variance of which is equal at all values of x. Thirdly, the distribution of possible y values at each x should have a normal distribution. Finally, all observations must be made independently of each other. This assumption can be met only by appropriate experimental design.

As for other parametric tests, data transformation may be required to achieve a linear relationship. In general, however, conversion of data to $\log_{10}$ or natural log (ln) is most likely to correct a problem of non-linearity. To check linearity, a scatter plot between y and x values can be generated by hand or, more accurately, using computer packages such as Microsoft Excel or Minitab. There are several other ways to check normality of data value distributions such as a histogram (looking for a symmetrical profile about the mean) or a normality plot in statistical packages. If the relationship between y and x is clearly not linear, transformation of data may help but would be a pointless exercise in cases where the data distribution curve contains distinct peaks or valleys (e.g. a bimodal distribution). Transformation may work if a curve decreases or increases uniformly from left to right. Transformations can be made on either y or x (or both) values, whichever works best. The method adopted depends on the exact shape of the curve. For example, taking the log value of one variable may help to linearize the data on an otherwise exponential curve. Obviously, the closer the correlation coefficient is to a value of 1 (typically 0.99) the better the evidence of a linear relationship between the two variables. An example of linear regression analysis and its presentation in graphical form is shown in Figure 4.2.

Most students by this stage will be familiar with the use of Microsoft Excel for basic graphical presentation. The package can also be used to apply a number of basic mathematical and statistical formulae (e.g. linear regression, descriptive statistics, t-tests, etc.) to aid data analysis. However, other packages exist that do a similar job.

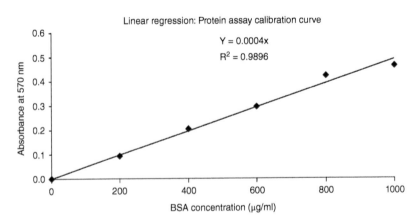

Figure 4.2 Linear regression in a protein assay calibration curve. Shown is an example of a typical protein assay curve plotted using Microsoft Excel with linear regression applied. The assay used the bicinchoninic acid (BCA) reagent and various concentrations of protein standard (bovine serum albumin; BSA), as indicated. The absorbance of the protein following incubation with BCA reagent was recorded at a wavelength of 570 nm. In this case, a good fit was observed with a high correlation coefficient (R^2), indicating a reliable standard curve. The equation of the straight line can be used to directly calculate the protein concentration value (x) corresponding to a given absorbance measurement (y).

A more powerful statistical package (e.g. Minitab) is recommended for more in-depth analysis, as it has a more comprehensive range of statistical tests and analyses, and a comprehensive help section. However, it may not be as effective as Excel or other software as a graphical package. As 'cells' of data can be readily transferred from one package to the other, combined use of the two packages using their respective strengths seems like a reasonable compromise. Information on using Minitab for data analysis is usually available in the '*Help*' facility on the statistical analysis package toolbar.

You should always remember that statistical analysis should be used to support your data by confirming its validity under rigorous and appropriate testing methods. It should not be used to try to get something out of a bad set of data – if the data are unreliable or inconsistent, it is better to repeat or redesign the experiment.

Uses of statistics

'If your experiment needs statistics you ought to have done a better experiment' – Ernest Rutherford

'Statistics are used much like a drunk uses a lamp post; for support, not illumination' – Vin Scully

4.5 Presentation, Structure and Organization of Data in Laboratory Reports

Effective presentation of your data is a key element in writing a good laboratory report. Developing these skills now will pay dividends later when you are analyzing more complex sets of data obtained from laboratory classes in later stages of your studies, from laboratory work on work placements or from your research project.

You will normally receive specific instructions on how to write the report from the lecturer in charge of the laboratory class. You are advised to follow any specific instructions given by your lecturer. However, the following hints and tips may also help you to produce a good report.

Structure

- Unless instructed otherwise by your tutor, a good report should typically be organized into sections containing an introduction, results and discussion/conclusions.
- Some reports may also require an abstract (summary), rationale, aims and/or a hypothesis, and a reference list citing sources of information used. Large reports may also benefit from a page indicating the main contents of the report and the page numbers where they can be found.
- Make sure that you include all of the sections requested in the assignment guidelines provided by your tutor.

Use of English

- A well-written report helps to give a good impression to the person assessing it. Thus, if you are writing by hand, try to make your writing as neat (legible) as possible and avoid untidiness.
- If your report is word processed, make sure that you check the spellings and consistency in font style, size and formatting of the text. You will often come across the situation in which spell check does not recognize a scientific word and suggests an alternative. Do not automatically assume that spell check is right; it is possible that the word is not on the spelling database. You can, of course, add the word to the spell check dictionary but make sure that you have spelt it correctly before doing so; otherwise, every subsequent use of the word will have the same error.
- It is convention in most areas of bioscience to use the past tense in the passive voice when describing methods and results in a report.
- Examples of typical scientific and grammatical writing errors are shown in Table 4.1.

4.5.1   Data Presentation

- Unless otherwise instructed by your tutor, graphs, histograms, schematic diagrams and images, and so on, should be referred to as Figure 1, 2, 3, and so on. Always remember to refer to the figure or table, and so on, at an appropriate point in the text of your report. For example, 'As shown in Figure 1, there was. . .'
- Each figure or table should have a clear descriptive heading. For example, 'Figure 1: Effects of organophosphates on acetylcholinesterase activity' clearly indicates the nature of the experiment (i.e. an enzyme assay in the presence and absence of organophosphates) from which the data shown in the graph have been obtained.
- Do *not* refer to graphs as graph 1, graph 2, and so on, unless specifically requested to do so. Similarly, tables should be referred to as Tables 1 and 2, and so on, again with a clear descriptive title.
- Another important feature of good data presentation is to have clearly annotated figures and tables. Thus, in a table, each column and row should be clearly labelled to indicate what is being shown.
- Figures containing graphs need clear labels on both axes (e.g. Absorbance and concentration could be the labels on the y and x axis of a calibration graph for a protein assay as shown in the earlier example of linear regression).
- If a quantitative value is presented (e.g. activity or concentration) make sure that the units are correctly defined (refer to Chapters 1 and 2 for correct expression of units of size and concentration). Where possible give an indication of the spread of data. For example, if several replicate measurements were made in class, you could estimate the standard deviation and express it in the form of error bars in the figure or table.

Examples of appropriately annotated tables and figures are shown in Table 4.2, Figures 4.2 and 4.3. Hints and tips on presenting microscopy images have already been discussed in Chapter 2. Errors in any of the above factors will lose you marks.

4.5.2   Content

It is essential that the work you present has not involved cheating. Two major forms of cheating are plagiarism and collusion. The former involves copying the work of others verbatim and presenting it as your own,

Table 4.1 *Typical mistakes in written scientific English.*

Incorrect phrases	Suitable alternative phrases
Add one gram of NaOH[a]. or We added one gram of NaOH.	One gram of NaOH was added.
Then add 10 ml of reagent A[a]. or Then we added 10 ml of reagent A.	Then, 10 ml of reagent A *were* added[b].
Bovine Serum Albumin (BSA)[c]	Bovine serum albumin (BSA)
Foetal Bovine Serum (FBS)[c]	Foetal bovine serum (FBS)
Sodium Chloride[c]	Sodium chloride
We homogenized the samples for. . . or Homogenize the samples[a] for. . .	The samples were homogenized for. . .
Samples were spun[d].	Samples were centrifuged.
Put samples in eppendorfs[d].	Samples were placed into 'Eppendorf' tubes or microfuge tubes
Samples were vortexed[d].	Samples were vortex mixed.
The experiment worked because we got positive results[e].	The experiment worked well, as the expected results were obtained.
We got negative results[e] with compound A.	No effect was observed with compound A. or Compound A had no effect.
Graph 1 shows. . .	Figure 1 shows. . .
We never saw any reaction.	No reaction was observed under the conditions specified.
Low temperature inhibits the enzyme a lot.	Enzyme activity was significantly reduced at lower temperatures.
I think this means. . .	This suggests. . . or This indicates. . .
The aims of this laboratory class are. . .	The aims of this laboratory class were. . .
A bacteria[f]	A bacterium
A mitochondria[f]	A mitochondrion
A phenomena[f]	A phenomenon
The culture media	The culture medium (if only one medium involved)
As you can see. . .	As can be seen. . .
It could of. . .	It could have. . .
It should of. . .	It should have. . .
I conclude that. . .	In conclusion. . .

[a] This is what you would see in a protocol, but you need to convert it to the past passive when writing a methods section for a report or thesis.

[b] Note the verb agreement in this case; the subject in this case is '*10* ml of reagent A'.

[c] Make sure that you use the upper and lowercase letters correctly in chemical and reagent names. The examples shown assume that the sentence starts with this item. If it appears in the middle of the sentence, all words should be in lower case.

[d] These are typical examples of laboratory jargon used in general conversation. However, a full description must be given in a report or thesis. For example, 'Eppendorf' is the name of a company which makes tubes, centrifuges, pipettes and a range of other products. It is often incorrectly spelt, as in this example.

[e] Statements like positive and negative results are either very vague or entirely meaningless. You need to be more precise in the description of your data.

[f] Do not use the plural form of these words incorrectly.

Table 4.2 *The effects of toxin exposure on the morphology of cultured neuronal cells (This is an example of how to effectively annotate a table of data).*

Incubation time (h)	Treatment	Axons	Extensions	Round cells	Flat cells
4	Control	32.35 ± 2.2	29 ± 1.1	16 ± 0.5	84 ± 0.5
	Toxin	13.2 ± 1.4*	29 ± 1.1	15 ± 0.6	85 ± 0.6
24	Control	35.9 ± 6.5	24 ± 0.8	19 ± 0.8	81 ± 0.8
	Toxin	18.5 ± 1.1*	28 ± 1.5	25 ± 2.0	75 ± 2.0

Cultured cells were incubated in the absence (control) and presence of toxin for 4 and 24 h, after which their morphological features were assessed by light microscopy. Shown are the average counts ± SEM for four independent experiments. Asterisks indicate where changes were deemed to be significantly different from the control ($p < 0.05$; students' t test). The data suggest that the toxin has an inhibitory effect on the number of axons produced but no effect on other aspects of cell morphology.

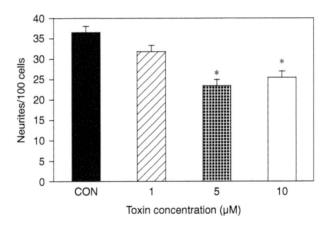

Figure 4.3 *Effects of toxin on neurite outgrowth (This is an example of a histogram plot with error bars). Neuronal cells were incubated with or without (CON) the concentrations of toxin indicated, and the effects of toxin measured on the outgrowth of cellular extensions ('neurites'). The experiment was repeated four times and the mean ± SEM calculated for each experimental condition. The plot was using Microsoft Excel and incorporated error bars for SEM values as shown. Asterisks indicate where there was a significant difference from the control ($p < 0.05$) using analysis of variance (ANOVA).*

while the latter may involve 2 or more students producing largely identical outputs when they were supposed to be independent reports. Most institutions have plagiarism software capable of detecting the degree of similarity with online material. You will no doubt receive clear guidance on plagiarism and how it is dealt with during your induction week.

Introduction
- If required, this should give information about the background to the study, leading into the aims of the laboratory class.
- It may require you do some background reading in textbooks; if this is the case, sources of information should be cited in the text and listed in full at the end of the report.
- This section could involve a mixture of present and past tenses, as appropriate.

Aims
- Clear, concise bullet points indicating the main experimental objectives of the laboratory class.
- This is best done using the past tense (i.e. The aims of this laboratory class/project were. . .)

Results

- Unless otherwise instructed, you should explain why each experiment was done.
- Give a simple commentary on the trends observed in each experiment, referring to each figure or table as appropriate.
- Present your tables and figures as indicated in the data presentation section above.
- Make sure that you contextualize all tables and figures in your commentary.

Discussion

- In this section, you should discuss the significance of the effects observed in your experiments (i.e. what does it *mean*). The depth required will vary from report to report, depending on the amount and complexity of the experiments involved.
- In longer reports (e.g. formal reports or project reports), you may be expected to relate your experimental findings to what is already known (i.e. published work), some of which you may have already covered in the introduction to the report. This is the opportunity to say whether or not your results agree with those reported by others and explain (discuss) any potential disagreements (e.g. on the basis of the experimental approach or materials used).
- Some mention of sources of error (e.g. sampling error, pipetting errors, inaccurate measurements, etc.) and possible ways in which the reproducibility of the data could be improved (e.g. performing more replicates and taking an average) would be a good idea in this section.
- Do not blame the protocol if everyone else got the experiment to work.

References

- Reference citation and reference list format should be consistent throughout.
- You should follow one recognized journal style. Numerous styles are used in books and journals that you will read; however, one of the most common formats is the Harvard system although there are numerous variations in use.
- The main emphasis should be on a *consistent* style, as this will look more professional. This system cites sources of information in the text by author surnames (if only one or two authors) or first author surname followed by *et al.*, then the year of publication if there are 3 or more authors. For example, 'Smith (1980), Smith and Jones (2001) or Jones *et al.* (2009)'.
- An alternative, equally acceptable form of referencing is the numerical system, which cites references by placing numbers in parentheses or superscripts according to the order in which they appear in the text. The full reference list should be included in the same numerical order at the end of the report.
- NEVER put full details of the reference in the text.
- Full details of the reference source should then be given in the reference list at the end of the report. For example, an acceptable reference list format could be:

4.5.3   Articles in Journals

Daleo, G.R., Piras, M.M. and Piras, R. (1974) The presence of phospholipids and diglyceride kinase in microtubules from different tissues. *Biochem. Biophys. Res. Commun.* 61, 1043–1050.

4.5.4   Articles in Books

Hatch, R.C. (1988) Poisons causing nervous stimulation or depression. In: *Veterinary Pharmacology and Therapeutics* (Booth, N.H. and McDonald, L.E., eds.), pp. 1053–1101, Iowa State University Press, Ames, Iowa.

4.5.5   Reference Management Software

There are numerous software packages available for managing and formatting references cited in your report (e.g. Reference Manager, Endnote, Mendeley, etc.). Such packages can be a useful aid, but you should still be aware of possible formatting glitches that can occur. You should still proofread references in terms of consistency of in-text citations (e.g. to ensure that author initials have not been included) and reference list formatting (e.g. to ensure correct use of upper- and lower-case letters in journal names).

4.6 Summary

Take special care in the way that you analyze, interpret and present your data. There is no point in carrying out the technical aspects of laboratory work competently only to present the data obtained badly in your report.

In terms of data analysis:

- When replicate experiments have been done, it is important that you perform appropriate statistical analysis, which will depend on the design of the experiment and the spread of the sample values obtained.
- If repeat measurements are made, you should at least perform basic descriptive statistics.
- If sample means are being compared, you need to determine the distribution and nature of the data sets before selecting an appropriate statistical test.
- Statistical analysis should be used only to support your conclusions, always bearing in mind that the limiting factor in statistical analysis is the quality of the data being analyzed.

In terms of data presentation:

- Always follow the agreed assignment guidelines.
- Present your work in a neat, tidy and professional manner using good scientific English.
- Write clearly and concisely, making sure that the appropriate sections are appropriately completed.
- Use clear headings, annotation and correct units on graphs and figures.
- If the experiment has not worked in your case, check the results of other students and discuss your own findings and what you should have expected, explaining possible reasons for the problems observed.
- If additional discussion is required, link your data to what is known in the literature and/or suggest possible sources of error and potential improvements.

Limitations of statistics

'I can prove anything by statistics except the truth'. – George Canning

'The individual source of the statistics may easily be the weakest link' – Josiah Stamp.

5

THE EXTRACTION AND CLARIFICATION OF BIOLOGICAL MATERIAL

5.1 Introduction

The basis of many biological experiments is the extraction and enrichment of cellular material. This usually requires two operational phases: (i) the extraction process and (ii) the clarification and enrichment of the component of interest. The method of extraction will depend upon the starting material; this could be abattoir material, plant tissue, eukaryotic or prokaryotic cell cultures. In addition to the choice of starting material, appropriate buffers must be used (see Chapter 1, Section 1.7) to ensure that the maximum amount of the desired component has been extracted.

5.2 Extraction

5.2.1 Proteins

Proteins are present within a cell at a high concentration, and they reside in a reducing atmosphere (little if no contact with oxygen). An extraction process immediately alters the normal environment of cellular protein by diluting the protein in a liquid, which is likely to be relatively oxygen rich due to the mechanical nature of the extraction process. Proteins are sensitive to heat (>40 °C) and changes in pH; therefore, the extraction of proteins from cells or tissue requires careful planning before the extraction process can be undertaken.

The method for the wholesale extraction of protein from cells or tissue will depend upon the starting material (animal, plant or bacteria).[1] In general, a buffer (0.1–0.25 M) cooled to 4 °C with the pH adjusted to 7.5 is a good starting point. The addition of other reagents to improve the recovery can be beneficial. Reducing agents such as dithiothreitol (DTT) or 2-mercaptoethanol (2-ME) can be added at 1–5 mM to help moderate the deleterious effects of the oxidizing extraction process. Chelating agents such as EDTA can be added to mop up any low concentrations of divalent heavy metal ions.[2]

In eukaryotic cells, hydrolytic enzymes are stored in the lysosomes of animal cells and the vacuole of plant cells. These compartmentalized enzymes do not normally encounter cellular proteins, but during the extraction process, the lysosomes (or vacuole) will be broken and they will mix with cellular proteins. There are two groups of hydrolytic enzymes that degrade proteins: peptidases (endopeptidase) which cleave the internal peptide bonds in a protein structure and peptidases (exopeptidases) which sequentially clip amino acids from a protein's amino or carboxy terminus. There are four major classes of peptidase (see Table 5.1): (i) serine peptidases possess a serine residue at the active site of the enzyme, (ii) sulphydryl peptidases possess a cysteine residue at their active site, (iii) acidic peptidases have a low pH optimum and (d) metallo peptidases have a requirement for metal ions. Other peptidases have been identified with other amino acids at the active site of these peptidases, e.g. glutamate and threonine.[3] The action of these enzymes can be controlled by the addition of commercially available peptidase inhibitor cocktails (see Table 5.1), thus improving the yield of intact proteins.

In addition to the additives outlined above, the extraction of protein from plant material can benefit from increasing the concentration of reducing agents up to 15 mM and including ascorbic acid (5–10 mM)[4] to help moderate the action of endogenous oxidases, the action of which can cause protein aggregation and denaturation.

Basic Bioscience Laboratory Techniques: A Pocket Guide, Second Edition. Philip L.R. Bonner and Alan J. Hargreaves.
© 2022 John Wiley & Sons Ltd. Published 2022 by John Wiley & Sons Ltd.

Table 5.1 *The major classes of peptidases and inhibitors.*

Class of peptidase	Popular inhibitor	Working range
Serine	Phenylmethylsulphonyl fluoride (PMSF)	0.1–1 mM
Sulphydryl	Iodoacetamide	10–100 μM
Acidic	Pepstatin	1 μM
Metallo	EDTA	1–10 mM

5.2.2 Deoxyribonucleic Acid

Deoxyribonucleic acid (DNA) is typically extracted from cells or tissue by disrupting the cell membrane with high salt and EDTA. The detergent sodium dodecyl sulphate (SDS) is then added to help unravel the structure of extracted proteins (see Chapter 6). Peptidase K (a non-specific peptidase) is added to digest the unwanted protein. After an overnight incubation, the DNA present in the extract can be isolated by either (i) repeated washing in high salt (6 M NaCl) or (ii) solvent precipitation (isopropanol) and repeated ethanol washing. The purity and quantity of the DNA extracted can be assessed using a spectrophotometer at 260 nm wavelength (see Section 3.8.1).

5.2.3 Lipids

Lipids can be extracted from cells or tissues with different combinations of solvents. Chloroform methanol and an acid (e.g. HCl or acetic acid) mixed in a ratio of 2000:1000:2 by volume is a popular choice, where 100 μl of the sample is mixed with 360 μl of the acidified chloroform:methanol. The addition of 120 μl of chloroform results in the formation of two phases. The chloroform-rich lower layer will contain virtually all the lipids, and the methanolic upper layer will contain the aqueous polar non-lipid components. Other solvent mixtures can be used including hexane/isopropanol and ethyl acetate/ethanol. The total lipids extracted can then be concentrated by drying (see Chapter 7, Section 7.4.4) and separated into different fractions using thin layer, normal phase or HiLic chromatography (see Sections 7.5 and 7.7).

KEY POINTS TO REMEMBER 5.1

When working with solvents, prefer to use a glass container because some plastics may not be compatible with the solvents. In addition, solvents may leach undesirable chemicals from the plastic. Solvents can be hazardous to health and the environment. Always work (and dispose of the used solvents) in accordance with the local health and safety procedures.

5.2.4 Organelles

The organelles of eukaryotic cells (e.g. nucleus, mitochondria, chloroplasts, peroxisomes, Golgi apparatus) possess a lipid bilayer structure like the plasma membrane surrounding the cell. To obtain intact organelles, an isotonic buffered solution (0.2–0.25 M sucrose) should be used. This will help to prevent the organelles from shrinking in a hypertonic solution or expanding (possibly bursting) in a hypotonic solution. The organelles can be harvested by differential centrifugation (see Section 5.6.1) and possibly further purified on a density gradient (see Section 5.6.2).

5.3 The Extraction Methods Used for Animal and Plant Tissue

There are many different extraction methods for eukaryotic and prokaryotic samples. To obtain the best yield from different starting tissues, certain techniques are more appropriate than others.

5.3.1 Mortar and Pestle

Small amounts of soft animal tissue can be conveniently extracted in buffer using a mortar and pestle.

- The mortar, pestle and buffer should be pre-cooled. Set the mortar on ice for the grinding process.
- Trim the fatty deposits from animal tissue and cut the tissue into small pieces using a scalpel or scissors.
- Place the tissue into the mortar on ice with a small amount of buffer and using the pestle, grind the tissue into a paste.
- Add more ice cold buffer if necessary (a final tissue to buffer ratio of 1:2 (w/v) or 1:4 (w/v) is usually sufficient).
- Strain the extract through four layers of muslin to remove large particles and then clarify the extract by centrifugation (see Section 5.5).

Acid-washed sand is sometimes added to the mortar to help in the extraction of plant material (or any other cells surrounded by a cell wall) as this can help disrupt the fibrous plant tissue. The processing of these extracts would be similar to the procedure outlined above for the extraction of animal tissue.

As an alternative method, the tissue (animal or plant) can be frozen in liquid nitrogen (take appropriate health and safety precautions when using liquid nitrogen) and then ground into powder prior to the addition of buffer.

5.3.2 Homogenizers and Tissue Grinders

As an alternative to a mortar and pestle, small amounts of animal tissue (or cells) can be disrupted using a homogenizer (see Figure 5.1). The pestle can be made from glass (Dounce homogenizer) or from Teflon (Potter–Elvehjem homogenizer), and they are manufactured to accommodate different volumes. The pestle can be plunged and rotated by hand or by a motor. Heat generated during the extraction process can be dissipated by surrounding the glass mortar with ice. This method of extraction will also work for cultured plant cells, but in general, it is not abrasive enough for most plant tissue.

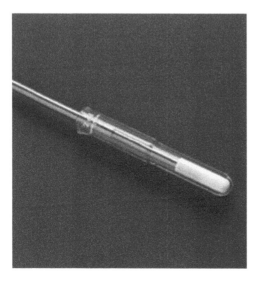

Figure 5.1 *A picture of a handheld Potter–Elvehjem (glass-Teflon) homogenizer. Taken from http://uk.vwr.com/app/catalog/Catalog?parent_class_id=9&parent_class_cd=60478.*

5.3.3 Blenders

Larger amounts of tissue, whether animal or plant in origin, can be extracted in a blender (e.g. Waring). The extraction vessel of a Waring blender (like a domestic blender) can be pre-cooled, and small fragments of tissue can be added onto the cutting blades at the bottom of the vessel. Enough cooled buffer should be added so that a vortex is generated when the blender is switched on. The extraction time can vary (5–20 seconds), but it should be interspersed with cooling phases (place the blender on ice for 60–120 seconds). After the blender has fragmented the tissue, the extract can be filtered through muslin (or cheesecloth) and then clarified by centrifugation (see Section 5.5).

A Polytron homogenizer (e.g. Ultraturax®) is another type of blender with a rotating blade at the end of a handheld probe. The probe can be placed into a buffered solution on ice containing the tissue. When the probe is switched on, the tissue is drawn through narrow gaps at the tip of the probe towards the rotating blades, resulting in efficient tissue fragmentation.

5.4 The Extraction Methods for Bacteria

Bacterial cells (diameter ≈ 0.5 μm) are approximately the same size as eukaryotic mitochondria. They are surrounded by a peptidoglycan cell wall made from polymers of sugars and protein. The presence of the cell wall and the relatively small size of bacteria make them difficult to break open by the homogenization techniques used for the extraction of animal and plant tissue outlined above.

Small volumes of bacterial cultures collected by centrifugation (13 000 $\times$ g for 20 min at 4 °C; see Section 5.5) can be broken open by incubating the bacterial suspension in a buffered solution of lysozyme.[5] This enzyme weakens the cell walls of bacteria by hydrolyzing the bonds that hold the sugar polymers together. A detergent (e.g. SDS) to disrupt the bacterial plasma membrane may be included to ensure the complete disruption of the bacterial cells. Large volumes of bacterial cultures can be collected into a pellet by centrifugation (see Section 5.5) and diluted into an extraction buffer before being placed into the chamber of a pressure cell (a thick steel cylinder). The piston of the pressure cell is placed into the chamber on top of the bacterial cell suspension. The loaded pressure cell is placed into a mechanical press and then subjected to pressure (10 000–40 000 psi). A valve at the base of the pressure cell allows the pressure to be released, forcing the bacterial cell suspension through a small aperture. The shattered bacteria in the extraction buffer can be collected into a beaker on ice. Most of the bacteria will be fragmented in this process, but it may be necessary to repeat the process to ensure complete extraction.

A sonicator provides a focused output of energy in the form of ultrasound waves (20 kHz) that can create holes in cell membranes, which results in cellular disruption. Sonicators can be used to fragment fragile cultured animal or plant cells, but bacterial cells may also need the presence of small beads (Ballatini beads) to knock holes in the bacterial cell wall and plasma membrane.

KEY POINTS TO REMEMBER 5.2 EXTRACTION

- Different combinations of extraction methods may prove more effective than a single method.
- The extraction process will generate heat, which will denature proteins. Always pre-cool the extraction buffer and equipment (if possible) before use and extract on ice or include cooling cycles (e.g. 15 seconds extraction followed by 30 seconds on ice).
- Re-extracting the pellet after the first centrifugation can improve the yield.
- During extraction, the contents of the lysosomes (animal) or vacuoles (plant) will be released into the extraction buffer. The internal pH of these hydrolytic organelles is approximately 5.5, which will lower the pH of the extraction buffer and make the final pH more attractive to the hydrolytic enzymes contained within these eukaryotic organelles. Check the pH (see Chapter 1) of the extract after extraction and adjust to the starting pH (see Chapter 1, Section 1.7) as necessary.

5.5   Clarification

After disrupting the tissue or cells of interest, a cell extract will contain a mixture of soluble and non-soluble (particulate) materials. In the extraction buffer, there will be soluble cellular components mixed with clusters of non-disrupted cells, different cellular organelles and/or fragments of cellular membranes and insoluble cellular components (e.g. starch from plant cells). The largest of these fragments and clumps of cells can be removed by filtering the extract through muslin (or cheesecloth), but the final clarification will require centrifugation.

5.5.1   Introduction to Centrifugation

In biosciences, there is often a need to separate the item of interest from other components; for example, blood samples contain both red and white cells, which can be separated from the blood plasma by centrifugation.

Any particle suspended in a liquid will sediment (or float) depending upon the size (and shape) of the particle, the viscosity of the medium and the density difference between the particle and the liquid. The rate of movement of a particle in a liquid is also influenced by gravity (the force of gravity (g) we experience; $g = 981$ cm s^{-2}) resulting in large dense particles sedimenting first. To separate particles in a liquid using these properties, it is often necessary to increase the applied gravitational force (g).

Gravitational force can be increased in a controlled environment using centrifugation in a machine called a centrifuge. Centrifuges are common instruments in any bioscience laboratory, providing a straightforward method for clarifying extracts. They produce a centrifugal force by spinning a sample about a central axis (see Figure 5.2). Depending on the type of centrifuge, the centrifugal force can be varied from a few hundred g to over half a million g by altering the rate of rotation. The relative centrifugal force (RCF) depends upon the speed of the rotor (see Section 5.5.3) in revolutions per minute (RPM) and the radius of the rotor (r, in mm).

Centrifuges and centrifuge rotors differ in their specifications (angle, radius, etc.) from one manufacturer to another. This means that setting the centrifuge to run using RPM may result in different gravitational forces being applied to an extract because the RPM does not take into account the shape or radius of the rotor that is placed onto the rotor spindle. To allow duplication of the clarification process between different laboratories, the RCF is preferable to RPM because it enables scientists to standardize the clarification conditions. Most modern centrifuges will have the option to run the centrifugation process in RPM or RCF (preferable). If the centrifuge does not include this facility, the data sheet for the rotor (search the manufacturers' website) will disclose the average radius or a tape measure can be used as a last resort. The worked example 5.1 shows how to convert RPM into RCF. Alternatively, if the appropriate measurements are known, a nomogram (see Figure 5.3) can be used to calculate the RCF.

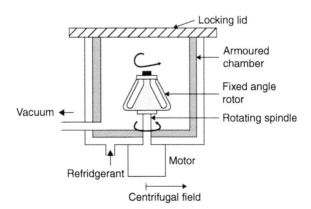

Figure 5.2 *A diagram of a basic centrifuge.*

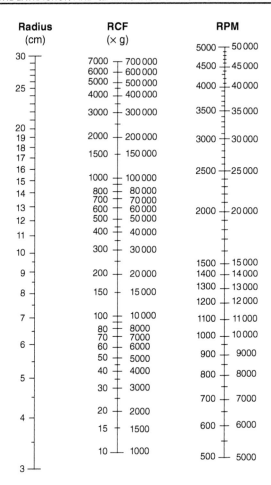

Figure 5.3 *A nomogram that can be used to calculate the RCF of a centrifuge rotor. To determine the maximum RCF, measure the radius (mm) from the centre of the centrifuge rotor to the furthest point on the centrifuge tube as it lies in the rotor while spinning. Draw a line from the column on the right of the nomogram (radius value) to the left-hand column (rotor speed in RPM). The RCF value is where the line crosses on the middle column, from which the value can be obtained.*

Worked Example 5.1 The Conversion of Revolutions per Minute (RPM) to Relative Centrifugal Force (RCF)

Relative centrifugal force (RCF) = $1.118\, r\, (RPM/1000)^2$
RPM = revolutions per minute
r = the radius of the rotor in mm

- Calculate the RCF in a centrifuge rotor spinning at 3000 RPM if the average radius of the rotor is 95 mm.

$$RCF = 1.118 \times 95 \times \left[\frac{3000}{1000}\right]^2$$

$$= 1.118 \times 95 \times (3)^2$$

$$= 956 \times g$$

What would the RCF be if the speed of the same centrifuge was increased to 6000 RPM with the same average radius of the rotor (95 mm)?

$$RCF = 1.118 \times 95 \times \left[\frac{6000}{1000}\right]^2$$
$$= 1.118 \times 95 \times (6)^2$$
$$= 3824 \times g$$

Note: There is not a linear relationship between RPM and RCF.

In this example, the speed of the centrifuge in RPM has been doubled but the RCF has increased four-fold. For this reason, it is accepted practice to quote the conditions for centrifugation in RCF using the average radius of the rotor rather than RPM. Information on the RCF, the temperature and the time of the centrifuge run allow the centrifugation conditions to be reproduced on different centrifuges which use rotors with varying diameters.

5.5.2 Centrifuges

Centrifuges (see Figure 5.2) are machines with an electrically driven central spindle, onto which a rotor is placed. For safety reasons, the spindle and rotor are contained within an armoured chamber with a locking lid. Centrifuges differ in their top speed, the size and volume of rotor they can accommodate and whether they are refrigerated or not.

Microcentrifuges (microfuges) are capable of rapidly accelerating small volumes (contained in 0.5-, 1.5- or 2.0-ml plastic tubes; called microfuge tubes to maximum speed of about 12–20 000 RPM (RCF > 10 000 × g).

Benchtop centrifuges or low-speed centrifuges can be used to harvest cells or large particles. They usually have a maximum speed of 3000–6000 RPM. The maximum RCF value will vary depending on the rotor used.

High-speed centrifuges are larger machines with a top speed of approximately 25 000 RPM (depending on the rotor, the RCF can be up to 75 000 × g). They are used to harvest protein precipitates and sub-cellular organelles. These machines need to be refrigerated to counter the frictional heat generated as air passes over the rotating surface of the rotor during the run. There are also benchtop versions of this type of centrifuge which are useful for smaller volumes of sample.

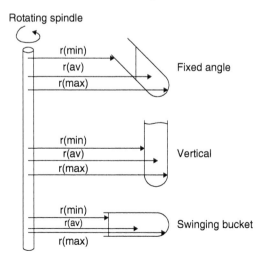

Figure 5.4 *The radius measurements in different centrifuge rotors. Minimum (r_{min}), average (r_{av}) and maximum (r_{max}) distances from the centre of the rotor.*

Ultracentrifuges are specialized machines that can operate at speeds up to 80 000 RPM (depending on the rotor, the RCF can be up to 350 000 × *g*). Slow acceleration and deceleration rates are usually available to prevent sample mixing. This is particularly important when a sample is placed onto a density gradient using a 'swinging bucket' rotor (see Figure 5.4). Ultracentrifuges require the chamber to be evacuated and refrigerated before the run can commence to higher speeds. They are used to sediment cellular membranes as well as DNA and RNA. Benchtop ultracentrifuges can be used to subject small sample volumes (0.2–5.0 ml) up to 650 000 × *g*.

Continuous (flow) centrifuges can be used for harvesting cells from large volumes of medium as the liquid flows continuously through a special rotor. This type of centrifuge is also used in the biotechnology and pharmaceutical industries to process large volumes of sample

5.5.3 Centrifuge Rotors

Centrifuge rotors can be divided into three different types: (i) fixed angle rotor, (ii) swinging bucket rotor and (iii) vertical rotor (see Figure 5.4). These different types of rotors are selected depending upon (i) the type of centrifugation required (differential, rate zonal or isopycnic), (ii) the volume of sample available and (iii) the RCF required to complete the experiment.

Microcentrifuges and analytical centrifuges are usually fitted with fixed angle rotors engineered from a solid block of aluminium, titanium, ceramic or carbon fibre. These rotors are a popular choice for routine extract clarification tasks. The rotors must have a smooth surface to reduce wind resistance and be conductive to dissipate the heat generated by movement through the air in the rotor chamber at high speeds. Routine checks for surface scratches or metal fatigue should be undertaken on a regular basis.

When a sample requires pelleting in a fixed angle rotor, the particles in the liquid move away from the centre of rotation throughout the whole centrifuge tube. They eventually contact the wall of the tube (see Figure 5.5b) and then start to slide down the wall of the tube to the furthest distance from the centre of rotation (r_{max}) forming a pellet. If the run time of the centrifugation is sufficient, the pellet will be located at r_{max}. Care should be taken when removing a sample after centrifugation in a fixed angle rotor to avoid disturbing the precipitated pellet. Fixed angle rotors come in a variety of different sizes to accommodate different volumes. Large rotors capable of accommodating larger volumes have a lower maximum speed limit and hence a lower maximum RCF. The K-factor of a rotor (supplied in the rotor data sheet) determines the pelleting efficiency of a rotor (a low K-factor indicates a higher pelleting efficiency), and this can be a useful parameter when comparing the use of different rotors. Efficient rotors have a high maximum RCF and a low sedimentation path length resulting in a low K-factor.

Benchtop centrifuges are routinely fitted with swinging bucket rotors. These swinging bucket rotors are useful for harvesting of cells from culture or processing blood samples (10–1000 ml). High speed centrifuges also have swinging bucket rotors for rate zonal or isopycnic centrifugation (up to 36 ml) (see Figure 5.5a and Sections 5.6.2 and 5.6.3).

The samples are loaded into buckets in the vertical position and, during a centrifugation run, the buckets move into the horizontal position. In a differential centrifugation experiment, all the particles in the liquid will be subjected to maximum RCF (see Figure 5.5b) and they will eventually pellet at the bottom of the centrifuge tube.

As vertical rotors have specialized applications, they are not commonly used. However, they are useful in isopycnic separations (see Figure 5.5a and Section 5.6.3) of DNA in caesium chloride gradients.

KEY POINTS TO REMEMBER 5.3

- Regular maintenance of a rotor will prolong its life and help prevent accidents.
- After a centrifugation run has finished, always rinse the rotor with water to remove any small spillages and then leave the rotor to dry upside down on tissue paper. Dislodge any debris from the rotor with a *soft* cloth or tissue, not a brush.

Check the inside of the centrifuge basin for small spillages and clean thoroughly.

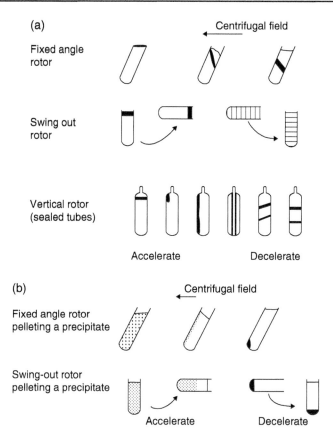

Figure 5.5 *What happens to particles during a centrifugation run: (a) gradient conditions and (b) pelleting conditions.*

5.5.4   Centrifuge Tubes

There are many different types of centrifuge tubes manufactured from a variety of materials to accommodate varied sample volumes. Before selecting a centrifuge tube, consult the data sheets from the manufacturers on the most appropriate tube to use and their chemical compatibility. Regular glass tubes should not be used above 3000 × g. Toughened Cortex™ glass can be used up to 25 000 × g, whereas polycarbonate (clear), polyethylene (clear), polypropylene (opaque) and polyallomer (opaque) can be used up to 100 000 × g. Thicker walled tubes may be required for speeds above 100 000 × g.

KEY POINTS TO REMEMBER 5.4

- Centrifuge tubes are used to contain the sample during a centrifugation run. A failure of a tube's integrity during a centrifuge run could result in the rotor becoming unbalanced, damaging the rotor spindle. Furthermore, hazardous or corrosive material could be released into the rotor and the centrifuge chamber.
- If a tube shows signs of 'wear and tear', it should be discarded.

5.6 Centrifugation Techniques

In the 1920s, T. Svedberg and J.W. Williams undertook pioneering work in centrifugation. In honour of that work, the sedimentation velocity of a molecule (or organelle) in a centrifuge is called the Svedberg constant (sedimentation coefficient) (S). The Svedberg constant (1×10^{-13} seconds) is a ratio of a particle's sedimentation velocity (ms^{-1}) to the applied acceleration. Larger molecules sediment more quickly than smaller molecules with the same density, and they have a larger S number. For example, eukaryotic ribosomes (80S) comprise a large (60S) and a small (40S) subunit. There are several different centrifugation techniques that can be used to clarify homogenates and enrich organelle preparations.

5.6.1 Differential Centrifugation

This form of centrifugation exploits the size and mass of a particle rather than the density of a particle to affect a separation. The pelleting of precipitates and the separation of eukaryotic organelles can be achieved using progressive increases in the RCF and the time this force is applied to particles. The different particles present will sediment in order of decreasing size and mass.

- A eukaryotic cell homogenate contains organelles with similar densities, but there are differences in their size (see Table 5.2). Differential centrifugation can be used to separate these particles by a stepwise increase in the RCF and the time that the RCF is applied to the extract (see Figure 5.6).
- The cell aggregates, individual whole cells and nuclei can be collected in a pellet by centrifugation at $2000 \times g$ for 2–5 min.
- The supernatant from this first step can be centrifuged at $20\,000 \times g$ for 30 min to pellet the mitochondria and a mixture of membranes from the lysosomes and endoplasmic reticulum.
- The supernatant from this step can then be centrifuged at $80\,000 \times g$ for 60 min to produce a pellet of mixed microsomal membranes consisting of membranes from the plasma membrane, the lysosomes, the endoplasmic reticulum and the Golgi body (also other organelles such as the peroxisomes).
- The supernatant can be subjected to centrifugation at $150\,000 \times g$ for 180 min to pellet the ribosomes.

The fractions produced at each step will not be pure, but they can be said to be enriched in an organelle. The purity of each fraction can be assessed by measuring the activity of enzymes (marker enzymes; see Table 5.2) known to be present only in the target organelle. Alternatively, the proteins in each fraction can be separated by electrophoresis under denaturing conditions and transferred onto nitrocellulose during Western blotting (see Chapter 6, Section 6.3.3). The blot can then be probed with antibodies to proteins from different organelles (see Table 5.2) to demonstrate enrichment or contamination of a centrifuge fraction.

Table 5.2 *The diameter and density of some eukaryotic organelles. (Based on Table 4.7 'Protein purification (2nd Ed)' P. Bonner, Taylor and Francis (2018)).*

Organelle	Diameter (μm)	Density (g cm^{-3})	Marker enzymes	Marker proteins
Nuclei	5–10	1.4	DNA dependent RNA polymerase	Lamin A
Mitochondria	1–2	1.19	Succinate dehydrogenase	Monoamine oxidase or cytochrome C
Lysosomes	0.2–2	1.12	Acid phosphatase	Lysosomal-associated membrane protein (LAMP 1 and 2)
Endoplasmic reticulum	(vesicles of variable diameter)	1.13–1.15	Glucose 6-phosphatase	Calreticulin
Golgi	(vesicles of variable diameter)	1.03–1.06	NADP$^+$ phosphatase	ATP binding cassette transporter 1 (ABCA-1)
Plasma membrane	(vesicles of variable diameter)	1.02–1.04	Na$^+$ K$^+$ ATPase	Sodium potassium ATPase

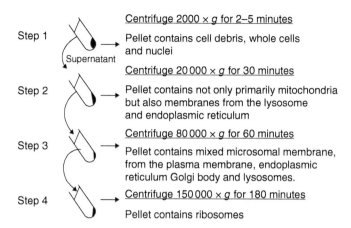

Figure 5.6 *Differential centrifugation of a mammalian tissue extract.*

5.6.2 Density Gradient Centrifugation

- Differential centrifugation of a eukaryotic cell extract will produce fractions enriched in an organelle. These fractions can be purified further by the density gradient centrifugation technique.
- A step gradient of sucrose (or other reagents such as Ficoll, Percoll or Nycodenz) can be generated by progressively layering sucrose solutions on top of each other (40% w/v sucrose at the bottom up to 10% w/v at the top of the tube).
- After differential centrifugation, the organelle-enriched fraction can be further purified by layering the fraction onto the sucrose step gradient and centrifuged in a swinging bucket rotor (see Figures 5.4 and 5.5) at 80 000 × g for 60 min.
- The membranes will migrate through each fraction until they are prevented from moving further by the density of the next layer of sucrose.
- At the end of the experiment, the interfaces of each layer are enriched in membranes of a similar density and these can be removed by careful pipetting.

5.6.3 Isopycnic Centrifugation (Equilibrium Density Gradient Centrifugation)

- As an alternative to density gradient centrifugation, the sample after differential centrifugation can be layered onto a continuous 20–70% (w/v) gradient of sucrose (or other reagents described above).
- During centrifugation, the components will migrate down the gradient, until they reach a point in the gradient where their density matches that of the sucrose.
- At the end of the experiment, the layers can be removed by careful pipetting or by puncturing the tubes and collecting fractions using a peristaltic pump.
- Vertical rotors are the most appropriate rotor for this technique but swinging bucket rotors can be used.

5.7 Points of Good Practice in Centrifugation

Prevention and regular maintenance are the best means to manage the risk of leaks. The correct use of the tubes, rotors and centrifuges will also help to prevent unnecessary damage. The manufacturer's website will contain help files giving guidelines on how to use their products correctly.

- *Never* use a centrifuge unless you have received the appropriate safety training.
- *Never* override any of the centrifuge's safety features.

- Log your use of the machine and provide information on the nature of your sample and how the operator can be contacted.
- Always check that the rotor is clean before use (rinse and clean if required; see Section 5.5.3).
- Switch the centrifuge on and place the rotor onto the spindle in the centrifuge bowl. Allow them both to cool to the operating temperature (the rotor may be pre-cooled in a fridge before use) before loading the samples.
- Only use the appropriate tubes for the intended rotor (see Section 5.5.4). Fill the tubes with distilled water and check for leaks prior to loading the samples.
- If the centrifuge tubes have been on ice, dry the tubes with tissue before placing them on the balance.
- *Never* add too much sample to the centrifuge tubes. Always fill the tubes to the level recommended by the manufacturers.
- Use a top pan balance to correctly balance pairs of tubes and remember to include the lids on the balance if they are to be used.
- Situate the balanced pairs of tubes in the rotor exactly *opposite* each other.
- When all the tubes have been loaded into the rotor, secure the lid of the rotor.[6] Close the lid of the centrifuge bowl (a locking sound is usually heard). If you are conducting a run in a high-speed analytical centrifuge, the vacuum pump will start.
- Programme the centrifuge according to the manufacturer's instructions and press the start button.
- It is advisable to stay with the machine until the required RCF has been reached. If the programme is halted and the speed returns to zero, open the bowl and check that there is no leakage into the rotor or centrifuge bowl. If a leak has occurred, clean the spillage in the bowl and rotor. Change the damaged tube and rebalance the centrifuge tubes.
- When the centrifuge is operating at the required RCF value, it can be left until the run time has been completed.
- Irrespective of the sample volume or tube size, if the samples have been loaded into a fixed angle rotor, the precipitate will be at the furthest point from the centre of the rotating spindle. If the sample has been loaded into a swing angle rotor, the pellet will be at the bottom of the centrifuge tube (see Figure 5.5).
- With smaller volumes in microfuge tubes (0.5–2.0 ml), a convenient way to identify where the pellet will be located after centrifugation in a fixed angle rotor is to ensure that the lid hinges of the microfuge tube are located towards the outside of the rotor. The required pellet will be located in the microfuge tube below the lid hinge.
- Remember even if you cannot see a pellet, dust particles and bacteria will be located at this point after centrifugation. It is important to carefully remove the sample tubes from the rotor without disturbing the pellet.
- Always pipette the supernatant from the tube at the point furthest away from the pellet (see Figure 5.5b).
- If the pellet is 'soft' (i.e. the pellet slides down the inner wall of the tube surface into the supernatant) or has been accidently disturbed, put the samples back into the centrifuge and start the centrifuge process again. You can increase the RCF and use the same length of time or use the same RCF for a longer period.
- After the centrifugation run has finished, *always* rinse the rotor with water to remove any spillages and then dry on tissue. Dislodge any debris with a *soft* cloth.
- To comply with health and safety procedures (and to ensure the machine's longevity), spillages must be reported and removed prior to the next centrifugation run.

5.8 Summary

- The cellular components of biological samples can be extracted by several different methods.
- To improve the yield of the component of interest, attention must be paid to the temperature, pH and components of the extraction liquid.
- Extracts can be clarified by increasing the RCF applied to the sample.
- There are many different types of centrifuges for different applications.
- In differential centrifugation, the cellular components can be enriched by progressively increasing the RCF and the duration of the centrifuge run.
- Organelles can be further enriched using density gradient centrifugations.

Notes

1 If a specific protein is to be extracted, the conditions necessary to ensure good yields of this protein can be determined in small-scale experiments.

2 The heavy metal ions in solution will bind covalently to reduced cysteine residues in the structure of any protein. This may alter the protein's conformation and/or biological activity.

3 The MERHOPS database (https://www.ebi.ac.uk/merops) contains a comprehensive list of identified peptidases which are now grouped into families according to their sequence similarity.

4 Ascorbic acid is regarded as an oxygen scavenger when the conversion to dehydroascorbate is catalysed by a divalent heavy metal ion, e.g. Hg^{2+}.

5 Another enzyme called lyticase can be used to disrupt yeast cells.

6 Some models of microcentrifuges do not have lids, and some models are fitted with a rotor lid to prevent excessive noise.

6

ELECTROPHORESIS OF PROTEINS AND NUCLEIC ACIDS

6.1   General Introduction

Electrophoresis is a technique that is used to separate charged molecules in an inert support matrix under the influence of an electric field. The most popular inert support for the electrophoresis of proteins is polyacrylamide (see Chapter 6, Section 6.2) and for the electrophoresis of nucleic acids, it is agarose (see Chapter 6, Section 6.4). When an electric field is applied, different biological molecules will move at different rates and the components of a mixture can be separated from each other. Electrophoresis has numerous applications in bioscience research, including the separation of proteins and nucleic acids, which will be the focus of this chapter. It is typically used in the analysis of complex mixtures (e.g. cell and tissue extracts) or in the monitoring of purity of isolated macromolecules (e.g. proteins, DNA, and so on).

The electrophoretic separation of any molecule depends on opposing forces of propulsion and resistance, as outlined below:

Propulsion is driven by:
a. Charge – negatively charged molecules (anions) migrate towards the anode (+) and positively charged molecules (cations) migrate towards the cathode (−).
b. Field strength – higher voltage promotes faster migration up to a certain limit, when overheating can occur, which will distort the electrophoretic migration pattern and could damage the resultant gel.

Resistance to migration is dependent on:
a. The size and shape of the biomolecules being separated – larger molecules migrate through a solution or matrix at a slower rate than smaller (lower relative mass) molecules as they experience greater frictional drag. Similarly, bulky molecules will experience more resistance than molecules with a linear shape.
b. The size of the pores in the support matrix – smaller pores will retard the rate of migration more than larger pores.
c. The viscosity of the buffer can slow the progress of molecules in electrophoresis.

Thus, the electrophoretic migration of individual biomolecules is dependent on the balance of the propulsion and resistance forces that they experience.

6.2   Separation of Protein Mixtures by Gel Electrophoresis

The most commonly used method for separation of proteins is polyacrylamide gel electrophoresis (PAGE) in the presence of sodium dodecyl sulphate (SDS) commonly known as SDS-PAGE. This technique is typically used to estimate the molecular weight of a protein by comparison of its electrophoretic migration with that of protein standards of known molecular weight which are used to construct a calibration graph (discussed in Section 6.2.3). The key steps in sample preparation, electrophoresis, staining and data analysis for SDS-PAGE and a brief overview of other application of electrophoresis in protein separation are outlined in the following sections.

Note: Electrophoresis involves the use of reagents that are potential carcinogens and teratogens. For example, the acrylamide, which is used to form the separation gel, is highly toxic in its monomer form either in solution or as a powder. Always make sure that you handle and dispose of such reagents according to local health and safety guidelines.

Basic Bioscience Laboratory Techniques: A Pocket Guide, Second Edition. Philip L.R. Bonner and Alan J. Hargreaves.
© 2022 John Wiley & Sons Ltd. Published 2022 by John Wiley & Sons Ltd.

6.2.1 Preparation of Electrophoresis Gels

Proteins are normally separated by electrophoresis in a three-dimensional cross-linked gel matrix formed from the polymerization of monomeric acrylamide in the presence of the cross-linking molecule *bis*-acrylamide. Typical ratios of acrylamide to *bis*-acrylamide are 29 : 1 and 37·5 : 1, and attention should be given to the fact that proteins will separate differently if changing from one cross-linker ratio to another in gel preparation. Any given protein might be expected to migrate more quickly at the lower cross-linker ratio (37·5 : 1) than at the higher one (29 : 1).

Nowadays, acrylamide is typically supplied as a prepared stock solution containing a mixture of acrylamide and *bis*-acrylamide in aqueous solution of either 30 or 40% (w/v) total acrylamide. This stock solution and distilled water are then mixed in various proportions with fixed amounts of Tris buffer and SDS, together with catalysts, to produce resolving gels of the required concentration for best resolution of the protein mixture to be separated. Typical recipes are shown in Table 6.1 for the preparation of resolving gels containing 7.5, 10 or 12% (w/v) polyacrylamide. Although a single resolving gel system can be used, better resolution of protein bands is achieved using a resolving gel (for protein separation) overlaid with a stacking gel (which allows rapid entry and accumulation of proteins at the resolving gel interface).

Manipulation of the polyacrylamide concentration determines the effective molecular weight separation range, over which good resolution can be obtained. The amounts of water and acrylamide solution would need to be adjusted if changing from one acrylamide stock solution to another, in order to get the same final concentration of acrylamide.

After mixing the main ingredients, the initiator (ammonium persulphate) and catalyst (TEMED) are added and immediately mixed. The mixture is then poured in between two glass (or plastic) plates separated by spacers (1–2 mm wide), sealed at the bottom and mounted on a gel assembly stand (Figure 6.1). The mixture is overlaid with distilled water or water saturated solvent (e.g. isobutanol) to provide a flat surface for the gel by removing the meniscus. The overlay solution also prevents direct contact with air, which inhibits polymerization.

Table 6.1 *Composition of resolving gels and stacking gels for SDS-PAGE.*

	Polyacrylamide concentration (w/v)* all volumes are in ml		
	7.5%	10%	12%
(a) Resolving gel (10 ml)			
Effective range of resolution (kDa)	30–300	15–150	5–80
40% (w/v) acrylamide stock solution (29 : 1) (comprising acrylamide and *N,N'*-methylene *bis*-acrylamide in a mass ratio of 37 5 : 1 or 29 : 1)	1.9	2.5	3.0
1.5 M Tris buffer, pH 8.8	2.5	2.5	2.5
10% (w/v) SDS	0.1	0.1	0.1
Distilled water	5.5	4.9	4.4
Mix thoroughly then add catalysts in sequence:			
10% (w/v) ammonium persulphate (AMPS)	0.10	0.10	0.10
N,N,N',N'-tetramethylethylene diamine (TEMED)	0.01	0.01	0.01
Mix well, pour and overlay with water or solvent until polymerized			
(b) Stacking gel (10 ml)	**4% (w/v)**		
40% (w/v) acrylamide stock solution (29 : 1)	1.0		
0.5 M Tris pH 6.8	2.5		
10% (w/v) SDS	0.1		
Distilled water	6.4		
Mix thoroughly then add catalysts in sequence:			
AMPS	0.10		
TEMED	0.04		

Mix well, pour onto the resolving gel surface and insert comb to form sample wells.

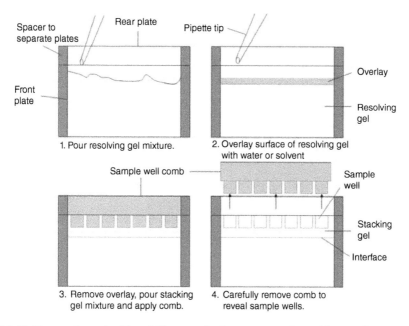

Figure 6.1 *Making a polyacrylamide gel. The type of gel most commonly used for protein separations is a dual gel system in which the resolving gel (which separates the proteins by size) is overlaid with a stacking gel (which allows proteins to enter the gel and accumulate at the interface with the resolving gel) of lower pH and acrylamide concentration. Shown are schematic representations of the major steps in gel formation. The stacking gel can be omitted but the resolution and sharpness of polypeptide bands are much better when a stacking gel is used. In such cases, the gel mould is filled with resolving gel mixture and the comb applied directly into it.*

After polymerization of the resolving gel, which may take one to two hours, the isobutanol (or water) layer is removed, the resolving gel surface is rinsed with distilled water and the stacking gel is mixed and applied. Once again, the catalysts are added and thoroughly mixed into the acrylamide solution just prior to pouring it into the gel assembly unit. As can be seen for the recipe in Table 6.1, the stacking is at a lower pH (6.8) and the polyacrylamide concentration (4% (w/v)) is lower than that of the resolving gel. This facilitates the rapid entry of all proteins into the gel as electrophoresis commences and their subsequent concentration at the interface between the stacking and resolving gels prior to entry into the latter. The wells of the stacking gel are produced by the insertion into the freshly poured stacking gel mixture of a sample well 'comb', the teeth of which form the mould that shapes the wells into which samples are loaded (Figure 6.1). If you are pouring your own gels, a stacking gel depth of about 1 cm is ideal. If you are attending an SDS-PAGE practical class of three hours or less, you may be provided with a precast polyacrylamide gel. In this case, you only need worry about following the instructions to open it; remove the comb carefully, to preserve the sample wells, and take away the sealing strip at the bottom of the plate to facilitate a continuous flow of ions and finally rinse out the sample wells with running buffer as for the homemade gels (see above).

The prepared gel assembly containing the homemade or precast gel matrix is inserted into the gel holding assembly with the smaller of the two glass plates facing inwards. The exact way in which this is done will vary slightly from one make of apparatus to another, but it basically involves careful positioning of the shorter plate of the gel assembly (or the appropriate surface of the buffer dam) against a rubber sealing gasket on the electrode unit so that they can be easily snapped into position. Either another gel or a 'buffer dam' (i.e. a piece of plastic shaped like the polymerized gel assembly) is applied to the other side of the electrode unit. This will create a watertight sealed inner chamber. Once assembled with a gel on both sides, the inner chamber is filled to a level below the top of the larger glass plate but above the top of the lower glass plate of the gel assembly. Two typical types of apparatus are shown schematically in Figure 6.2.

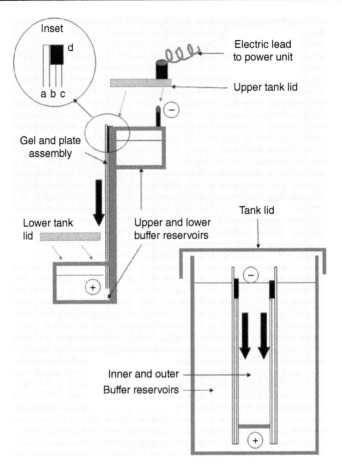

Figure 6.2 *Schematic diagrams of typical apparatus layout for SDS-PAGE. Shown are two of the major types of electrophoresis tank in cross section. The system in the top diagram employs an upper and lower buffer reservoir system with the cathode at the top. Some systems of this type allow a second gel to be attached to the other side of the upper buffer reservoir, with a lower buffer reservoir on the opposite side to the one shown. The system in the lower diagram employs an inner and outer buffer tank system with the electrodes arranged in such a way that protein migration is in the downward direction. The inset (top left) shows the alignment of the gel assembly with the rubber seal to create a watertight contact that prevents leakage. Indicated are the longer outer glass plate (a), the gel (b), the shorter inner glass plate (c) and the rubber seal (d). A similar type of seal is used in all systems. Note that only one terminal is viewed in the cross sections, but positive and negative terminals are present in both types of apparatus and these normally plug into sockets in the protective lids, connecting to the leads that are plugged into the power unit. Relatively few systems use horizontal (flatbed) gel electrophoresis, which is more common for nucleic acid electrophoresis (see example in Figure 6.5).*

6.2.2 Sample Preparation and Loading

In SDS-PAGE, proteins are separated on the basis of size differences. However, as discussed above, the electrophoretic migration of a molecule depends on several factors including its size, shape and charge. As each protein is unique in terms of its primary amino acid sequence, it follows that the net charge will vary from protein to protein depending on the balance of amino acids with positively (e.g. lysine, histidine and arginine) and negatively charged amino acid (aspartic and glutamic acid) side chains. In order to eliminate such charge differences, prior to their separation by SDS-PAGE, proteins are denatured (i.e. unfolded) by boiling them in sample preparation buffer (see Table 6.2) containing the anionic detergent SDS. Like other detergents, SDS is an amphipathic molecule, meaning that it has hydrophilic and hydrophobic components. In this case, the hydrophobic component (the alkyl chain) interacts non-covalently with hydrophobic amino acid side chains (e.g. valine), thus preventing the protein from precipitating in solution, whereas the hydrophilic portion (i.e. the sulphate group) projects outwards from the protein surface into the aqueous environment keeping the detergent-covered polypeptide in solution.

As the SDS binds uniformly along the polypeptide chains, at approximately every two amino acids, each protein will now have a constant charge: mass ratio. However, as many proteins also contain disulphide bonds between cysteine residues, their tertiary structure is not completely linearized unless a reducing agent is included in the sample preparation buffer. A common reducing reagent used in this case is 2-mercaptoethanol, which breaks down disulphide bridges, thus allowing the complete unfolding and linearization of the proteins (see Table 6.2). Proteins treated in this way become linearized polypeptide chains with constant charge:mass ratio, varying only in their chain length (i.e. mass).

After the stacking gel has polymerized (30–60 minutes), the sample well comb is carefully removed and the sample wells rinsed out with electrophoresis running buffer, typically Tris, glycine and SDS pH 8.3 (see Table 6.2). Glycerol (10% (w/v)) is added to the sample buffer to make it denser than the electrophoresis running buffer and a tracker dye (bromophenol blue) makes it more visible; thus, when samples are loaded directly through the running buffer using automatic pipettes, they can be seen sinking to the bottom of the wells (see the practical hints and tips box for the best way to load your samples). It is a good practice to apply a set of protein standards of known molecular weight with every gel. If the standards are pre-stained (i.e. they have a covalently bound dye attached to them), their migration can be monitored through the gel as electrophoresis proceeds.

Notes: Gel electrophoresis buffers are often prepared as master stocks (e.g. ×5 or ×10 strength) that are diluted to working strength with distilled water. The buffer compositions given are typical examples and represent typical (but not exclusive) working concentrations. You should also be aware that several variations exist, as indicated below:

In SDS-PAGE sample buffer, alternative reducing agents such as dithiothreitol can be used. Furthermore, in specific applications, a different pH may be used for both sample and running buffer.

In nucleic acid separations, sucrose and Ficoll are interchangeable and can be replaced by a similar concentration of glycerol. Various tracker dyes can be used and sometimes more than one dye is added to the sample buffer giving distinct dye fronts (DFs) reflecting their different molecular sizes. Other denaturing agents may replace NaOH (e.g. urea).

Table 6.2 *Examples of typical electrophoresis buffer compositions.*

Technique	Sample preparation/loading buffer	Electrophoresis running buffer
SDS-PAGE	2% (w/v) SDS in 62.5 mM Tris-base (pH 6.8), 10% (w/v) glycerol, 5% (v/v) 2-mercaptoethanol, 0.1% (w/v) bromophenol blue.	0.1% (w/v) SDS in 25 mM Tris-base (pH 8.3) and 192 mM glycine.
Non-denaturing PAGE	As for SDS PAGE except that SDS and 2-mercaptoethanol are excluded.	25 mM Tris-base (pH 8.3) and 192 mM glycine.
Non-denaturing DNA electrophoresis	Sucrose at 4% (w/v) in 90 mM Tris-base (pH 8.3), 80 mM boric acid, 2.6 mM EDTA (TBE buffer), to which one or more tracker dyes (e.g. bromophenol blue) can be added at 0.01% (w/v) or lower concentrations.	TBE buffer. Note that acetate can be used in place of borate to make 'TAE' buffer.
Denaturing DNA electrophoresis	50 mM NaOH, 1 mM EDTA, 2.5% (w/v) Ficoll (type 400), 0.01% (w/v) bromocresol green.	TAE buffer.

6.2.3   Running Electrophoresis Gels

As all proteins after the sample preparation step are now uniformly negatively charged, when the electric field is applied, they will move towards the anode (+). The larger proteins (i.e. longer polypeptide chains) will move more slowly than the smaller ones, as they experience greater levels of frictional drag as they move through the gel. Thus, smaller proteins will migrate more quickly and arrive at the end of the gel first. A negatively charged tracker dye, usually bromophenol blue, is added to the electrophoresis sample buffer. The blue dye moves ahead of all proteins forming the DF. When this reaches the end of the gel, the voltage is switched off and the gel removed carefully to avoid damaging the gel or the glass/plastic plates. Having dismantled the gel assembly, the gel is carefully lifted or floated off the plates into a staining solution that fixes and stains the proteins separated in the gel so that they can be visualized, and an image taken.

Proteins can be revealed by staining with Coomassie brilliant blue in a mixture of acetic acid (5–10% (v/v)) and methanol or ethanol (10–20% (v/v)), which can detect down to 0.1–0.2 µg of protein. A typical stained gel is shown in Figure 6.3. A more sensitive staining reagent containing fluorescent protein binding dyes (e.g. Sypro® Ruby) or silver chloride can be used to detect as little as 0.5 ng of protein. Alternatively, there are a number of recently developed eco-friendly Coomassie-based dyes that are very rapid, reduce the use of solvents in the fixation step and have a greater degree of sensitivity (e.g. CooBlue MAX or Bio-Safe Coomassie stain); these are colloidal Coomassie blue solutions that can detect as little as 10–20 ng of protein within one hour).

After staining, separated proteins appear as blue bands of the same width as the sample wells into which they were applied. Once the gels are stained, it is possible to measure the electrophoretic migration distance of each polypeptide band and compare them to the distance moved by the DF. An Rf value (retention factor or 'relative to the front') can then be calculated for each band, where:

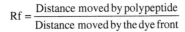

$$Rf = \frac{\text{Distance moved by polypeptide}}{\text{Distance moved by the dye front}}$$

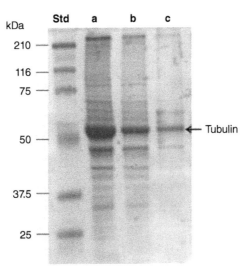

Figure 6.3 *A typical Coomassie blue stained SDS-PAGE gel. Proteins were separated by SDS-PAGE in a 7.5% (w/v) polyacrylamide gel overlaid with a 4% (w/v) polyacrylamide stacking gel. Shown, from left to right, are molecular weight standards (Std) and a protein extract at 10 µg (a), 3 µg (b) and 1 µg loadings. This extract was a preparation of microtubules purified from mammalian brain. The major protein (55 kDa) is tubulin and there are a number of less-abundant microtubule proteins. It can be clearly seen that the staining is quantitative over this protein loading range (i.e. the staining intensity is proportional to the amount of protein present). The molecular weights of the protein standards are shown to the left of the gel image. Note that although silver staining is more sensitive than staining with Coomassie blue, it is less quantitative. An example of a silver stained gel is shown in Figure 6.4.*

This will always be a ratio with no units and it will never be a value greater than one, as no protein molecule moves faster than the DF. If a set of protein standards of known molecular weight are electrophoresed in parallel to the samples, a calibration graph can be constructed between the $\log_{10}$ value of their molecular weight and their Rf values. This will be a linear negative slope, as indicated in the worked example below.

Worked Example 6.1 Molecular Weight Determination by SDS-PAGE

Complete the following tasks:

a. Measure the Rf values of the standard protein and sample bands in the schematic diagram of the SDS-PAGE gel provided.
b. Plot a graph between the $\log_{10}$ molecular weight (MW) and the corresponding Rf value for each protein standard.
c. Use the calibration graph to estimate the molecular weight of the sample band.

What to do:

a. To calculate the Rf value, you should measure the distances migrated from the start of the resolving gel (origin) to the centre of each band and divide this by the distance moved by the DF.
b. Put all of these values into a clearly annotated table as shown.
c. From the molecular weights given for the standards, work out the $\log_{10}$ values and put this in the corresponding column of the table, as shown.
d. It is then possible to plot a graph either by hand or using a software package such as Microsoft Excel, as shown below the table.
e. If plotting by hand, remember to stretch the log axis as much as possible. For example, as the log values range only between 4.45 and 5.31, the start and end of the x axis could be log 4 and 6, respectively (or even 4.4 and 5.4, as shown on the Excel plot). You should then draw the best fit straight line through the data points.
f. The advantage of using a package such as Excel is that the trend line can be fitted, and values for the equation for the straight line and the correlation coefficient can be displayed.

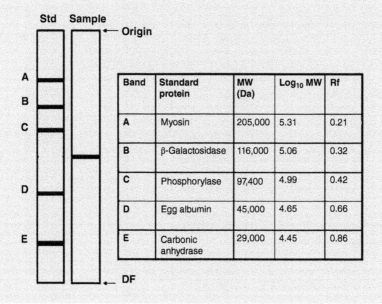

Band	Standard protein	MW (Da)	Log$_{10}$ MW	Rf
A	Myosin	205,000	5.31	0.21
B	β-Galactosidase	116,000	5.06	0.32
C	Phosphorylase	97,400	4.99	0.42
D	Egg albumin	45,000	4.65	0.66
E	Carbonic anhydrase	29,000	4.45	0.86

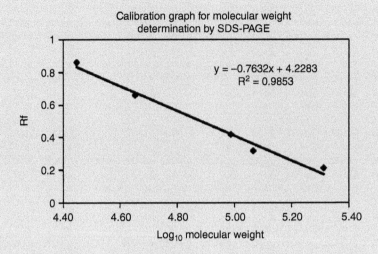

Calibration graph for molecular weight determination by SDS-PAGE

$y = -0.7632x + 4.2283$
$R^2 = 0.9853$

The high value for the correlation coefficient (> 0.98) indicates that the linear plot is good (see Chapter 4).

Now, do the calculation:

$$\frac{\text{Distance moved by the sample band}}{\text{Distance moved by the dye front}}$$

You should find a value of approximately **0.52.**

Remember it is a *ratio* and therefore has no units! This value can then be read off on your hand-drawn graph or substituted into the equation from the Excel plot shown above, giving a $\log_{10}$ molecular weight value of 4.858 884.

If you had drawn exactly the same graph by hand, the nearest you would get to this would be *4.86* (2 decimal places). Hand measurements cannot be more accurate than this.

You then need to calculate the antilog of this value in order to arrive at the estimated molecular weight.

Using the Excel equation value, the estimated molecular weight is *72 258* Da.

Using the 2 decimal place approximation, the molecular weight is *72 444* Da.

Thus the hand-drawn graph, done with precision, can give a reasonably accurate estimation of molecular weight.

Remember that this is only an estimate. Precise molecular weight determination would require more extensive sequence and chemical analyses. Repeating measurements on different gels and samples would also improve accuracy.

Student Exercise

Using the same calibration curve and equation, estimate the molecular weight in Daltons of polypeptides migrating on SDS-PAGE with Rf values of 0.48 and 0.72.

PRACTICAL HINTS AND TIPS FOR SDS-PAGE

Handling of gels and reagents: Gloves should be worn at all stages where reagents and gels are handled, not only for personal protection but also to avoid unnecessary introduction of keratins (skin proteins) into the sample or running buffers, and so on.

Sample preparation:

- Sample buffer usually contains volatile reducing agents such as 2-mercaptoethanol, which will expand on heating.
- To avoid release of pungent vapours into the atmosphere, you should place a clamp over the lid of the microfuge tube containing the sample during the high-temperature incubation step in order to prevent the tube lid popping off.
- As an additional precaution, if practical, samples could be boiled and loaded into gels in a fume cupboard (if available).
- It is safer to use a heating block to prepare samples for electrophoresis, as a water bath can run dry if left unattended.

Gel preparation:

- When preparing your own gels, make sure that all surfaces coming into contact with acrylamide are clean.
- Plates, combs and spacers should be cleaned with 70% (v/v) ethanol in order to remove traces of grease that might interfere with polymerization.
- It is becoming increasingly popular to use commercially available precast polyacrylamide gels. These are typically sealed at the bottom with a plastic strip. Failure to remove the sealing strip before turning on the electrical supply will prevent electrophoresis, as it will present a barrier to the flow of electrolyte.

Sample application:

- Careful pipetting is essential (see Chapter 1).
- When applying the sample into the sample wells, place the tip to the centre of the well avoiding contact with the sides, preferably using gel loading tips which have a thin extension that fits comfortably between the glass plates.
- Do not puncture the stacking gel with the tip as such damage may distort the gel pattern.
- Carefully eject the sample to the second position of the pipette barrel (see Chapter 1), keeping the barrel depressed until the tip has been removed from the sample well altogether; this will avoid suction of the sample back up into the pipette.
- The sample can be seen sinking to the bottom of the well due to the presence of glycerol and bromophenol blue.
- Precast gels usually have a template imprinted on them to outline the shape of the wells, thus facilitating the sample loading process.
- However, as is normally the case with manual gel casting, if you cannot see the outline of the sample wells when they are filled with running buffer, consider using a permanent marker pen to draw a point on the glass plate near the centre of each well, or to outline the bottom of each well, before assembling the electrode unit or before rinsing out with running buffer.

Vigilance during electrophoresis:

- Make sure that terminals and electrical leads are clean, dry and connected correctly to the apparatus and the power pack. If you plug in the leads the wrong way, your proteins will migrate upwards into the buffer tank instead of migrating downwards into the gel.
- The DFs of gels of the same polyacrylamide composition electrophoresed in parallel in the same apparatus should migrate at the same speed, giving a straight line. However, electrolyte leakage or low levels of electrolyte can prevent this. A big difference in DF migration or abnormal DFs can be indicative of uneven electrolyte flow.
- Always check at regular intervals that the buffer level in the inner (or upper) chamber is above that of the lower glass plate and that the DF on each gel is straight in appearance.
- If the running buffer levels fall too low, there will be no continuity between the electrolyte in the two chambers of the electrophoresis apparatus, the gel, and the electrophoretic separation will come to a halt.
- If topping up is required, switch off the power supply before removing the lid.
- Uneven leakage from gels being electrophoresed in parallel in the same apparatus will be evident as noticeable differences in the progress of the DFs of the different gels.
- Localized differences in electrolyte flow along the same gel can also occur, for example, due to the electrophoresis apparatus being placed on an uneven surface and a low level of electrolyte in the chamber. This typically results in a wavy appearance of the DF.

Gel removal and handling:

- *Switch off* the power before dismantling the apparatus.
- In order to avoid cuts or more serious injury, *never* use a scalpel or scissors to separate the plates from each other after electrophoresis is complete. This could result in damage to the glass plate, airborne chips of glass or scalpel blades.
- Always use either a plastic wedge that fits snugly between the plates or a thin blunt instrument (e.g. a spatula) to carefully separate the plates for gel removal.
- The gel can be lifted by gently picking it up, with gloved hands, via the two corners along the length of the bottom of the gel. Alternatively, it can be floated off the glass/plastic plate into the fixing or staining solution. Gentle agitation of the gel in the fixing solution usually facilitates rapid separation of the polyacrylamide gel from the plate. This can then be removed to allow the staining solution to penetrate the gel surface from both sides of the gel face.

Data analysis:

- As shown in the worked example, always plot $\log_{10}$ of the molecular weights of the standards, as the exponential curve produced in a plot of actual molecular weight against Rf, would lead to a less accurate estimation of molecular weight of a sample protein.
- In the analysis step, the distance moved should always be measured between the same reference points. For example, the point of origin should be the start of the stacking gel and the distance moved should be to the middle of the polypeptide band. In the event that a stacking gel is not used, the point of origin would be the bottom of the sample well.
- Bear in mind that inaccuracies could result especially from manual measurements using a ruler, for which accuracy would be 0.5 mm at best.
- Repeated measurements on different gels would help to improve the accuracy.

Interpretation:

- When asked to tentatively identify a protein based on its size, *do not* assume that a protein in the sample to be analyzed having the molecular weight of one of the standards provided is necessarily the same protein as the one in the standard.
- Standards are provided simply to allow the construction of a calibration curve for the estimation of molecular weights.
- SDS-PAGE will only give an approximation of the molecular weight of a polypeptide. An accurate isotopic relative molecular mass determination would require a more in-depth analysis (e.g. mass spectrometry).
- If a protein has a subunit structure, the subunits will be separated from each other using SDS-PAGE. Thus, a protein that exists in its native form as a homodimer (that is, two identical subunits) with a total molecular weight of 100 kDa, which could be determined by size exclusion chromatography (see Chapter 7), would migrate as two polypeptides of approximately 50 kDa each on SDS-PAGE.
- It should also be noted that not all proteins migrate in a log-linear fashion. For example, anomalous migration is often observed with glycoproteins due to the increased frictional drag caused by covalently linked oligopeptide chains projecting out from the linearized polypeptide chain. This can result in anomalous molecular weight determinations using SDS-PAGE and size exclusion chromatography.

6.3 Other Electrophoretic Techniques Applied to Proteins

A number of other applications involving the use of gel electrophoresis in protein separation are outlined in the following sections. The reader is referred to the reference list for further reading.

6.3.1 Non-Denaturing Gel Electrophoresis

In this approach, the protein sample is *not* denatured by boiling in SDS and reducing agent or by treatment with other denaturing agents such as urea. Instead, it is applied directly to the gel (polyacrylamide or agarose) and separated according to a combination of size, shape and net charge. An advantage of this technique is that proteins are in a more 'native' conformation, retaining their ability to form complexes with other proteins. After detection, spots can be eluted and separated under denaturing conditions by SDS-PAGE in order to analyze further the components of any protein complexes.

In solution, most proteins approximate to a globular structure, which is driven by the internalization of amino acids with hydrophobic side chains (e.g. leucine and tryptophan), which are accommodated in the core of the protein's structure to escape any contact with water molecules. This leaves the side chains of polar and charged amino acids (hydrophilic) to interact with water in the surrounding environment, thus keeping the protein in solution.

All proteins act as zwitterions because they contain amino acids with positively charged functional groups (e.g. lysine and arginine) and negatively charged functional groups (aspartate and glutamate). These charged groups are described as weak acid charges, meaning that the charge on the functional group is pH dependent (see Chapter 7 and Figure 6.5). At a physiological pH of 7.0, the surface of proteins is covered in both positive and negative charges. The positive and negative charge distribution for each protein will be different and depends upon the location of the charged amino acids in the primary structure of the protein (which is dictated by the primary amino acid sequence).

To impart enough charge to promote movement of a protein in an electric field, the pH of the electrophoresis buffer needs to be 0.5–1.0 pH units above a protein's isoelectric point (pI) to give it sufficient negative charge to move towards the anode (+ve). Conversely, the pH of the electrophoresis buffer needs to be 0.5–1.0 pH units below a protein's pI to provide sufficient positive charge to allow the protein to move towards the cathode (−ve).

Since more than 70% of proteins have a pI between 4.0 and 7.0, in an electrophoresis experiment using a buffer with a high pH value (e.g. 100 mM Tris/HCl pH 8.5) will result in most of the proteins having an overall negative charge, directing the movement of the majority of proteins towards the anode (+ve).

Electrolysis is the movement of a charged molecule in solution under the influence of an electric field. The charged molecules will migrate to and contact an oppositely charged electrode. In the case of metal ions, e.g. Cu^{2+}, this is not a problem and their integrity will be retained, but if the charged molecule is a protein, it would be destroyed on contact with the electrode by the heat at the electrode's surface. Also, in electrolysis, there is no resolution of a complex protein mixture; all the proteins in the mixture will migrate to an opposite charged electrode with little separation. Electrophoresis involves a simple alteration to electrolysis by the inclusion of an inert support between the two electrodes and terminating the experiment before the macromolecules are destroyed.

Polyacrylamide can be used to provide the inert support for the electrophoresis of proteins based on their molecular weight in the presence of the denaturing detergent SDS and also in the absence of SDS where the separation is based upon the size, shape and charge of the proteins in the mixture applied. PAGE in the absence of SDS is called "non-denaturing" PAGE or sometimes referred to as "native" PAGE. It uses similar experimental conditions as SDS-PAGE except that SDS and a reducing agent are not present either in buffers (see Table 6.2) or in the gel mixtures shown, for example, in Table 6.1.

In non-denaturing PAGE, the proteins migrate through the porous network of polyacrylamide towards the oppositely charged electrode in their native three-dimensional conformations. The proteins are applied to the gel and subjected to electrophoresis in their native conformations. The progress of a protein towards an electrode of opposite charge will depend upon two factors: (a) the overall surface charge on the protein, which is pH dependent (the exact surface charge at any pH is determined by the amino acid composition of the protein and any post-translational modifications, e.g. phosphorylation) and (b) the protein's hydrodynamic volume, with small compact structures migrating quicker than large disperse structures (changes to the conformation of the protein will also alter the rate of migration, e.g. the formation of aggregates and post-translational modifications such as glycosylation). For example, if two proteins with a similar Mr of 30 000 (a) protein X (containing 4 aspartate and glutamate amino acids) and (b) protein Y (containing 16 aspartate and glutamate amino acids) were subjected to non-denaturing PAGE at a pH above 7.0, protein Y would migrate further in the gel. Their masses are the same but the relative charge on protein Y is four times larger than protein X; therefore, in non-denaturing PAGE, protein Y will migrate further in the gel because protein Y is more electronegative. An example of the use of non-denaturing PAGE is shown in Figure 6.4. Storage globulins (*Vicia faba*) were isolated from sown seeds at different days after the onset of germination. The isolated globulins have similar molecular masses, but they show increased electronegativity due to the loss of amide nitrogen (i.e. conversion of asparagine and glutamine to aspartate and glutamate).

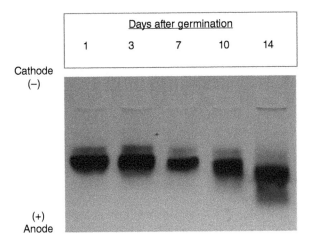

Figure 6.4 *Storage proteins isolated from Vicia faba seeds at different stages of germination. Storage proteins were extracted from Vicia faba seeds at different stages of germination, as indicated, and subjected to non-denaturing PAGE, after which they were stained with Coomassie brilliant blue. Note the increased migration of storage protein towards the anode caused by increased electronegativity is due to progressive deamidation of glutamine residues, which are converted to glutamate. Source: Adapted from: Bonner, P.L.R. (2018) "Protein purification: the basics" (2nd ed). Taylor and Francis, UK.*

6.3.1.1   Other Applications of Non-Denaturing PAGE

6.3.1.1.1   In Situ Detection of Enzyme Activity

If the temperature of the electrophoresis experiment is controlled by either circulating cold water or by placing the electrophoresis set up in an ice-bath, at the end of the experiment, the proteins in the polyacrylamide gel will be biologically active.

There are many published methods which allow for the detection of enzymes in the polyacrylamide gel (*in situ*) after the electrophoresis has been completed (see Table 6.3 for selected examples). The enzyme of interest can be identified post experimentally by placing the gel in a chamber or plastic bag containing a buffer solution with a chromogenic substrate for the enzyme of interest and incubating at an appropriate temperature (typically 37 °C) until a colour appears, representing the reaction product, in the area of the gel containing the enzyme of interest. Alternatively, a chromogenic substrate can be included in a solution containing agarose which can then be layered over the acrylamide gel, and subsequently incubated at 37 °C. As the chromogenic substrate is held near to the surface of the non-denaturing gel, facilitating an area of elevated substrate concentration, coloured reaction product will appear in the areas of the gel corresponding to the position of the target enzyme in the polyacrylamide gel. Peptidase, dehydrogenase and kinase activity assays are among the most popular methods to detect the presence of enzymes in a biological mixture. Non-denaturing PAGE is also useful to detect the presence of isoenzymes (two or more enzymes with the same catalytic function but with different primary structures) in a biological mixture. Isoenzymes can include families of enzymes with the same catalytic function transcribed from different genes or one enzyme with or without a post-translational modification, e.g. phosphorylation. Small charge differences will lead to altered electrophoretic mobility in non-denaturing PAGE.

If, after purifying a protein, a homogenous band is visualized using SDS-PAGE (separation on size) and the same preparation shows a homogenous band after non-denaturing PAGE (separation on hydrodynamic volume and charge), this can be taken as a good indication of a high level of the purity of the fraction.

The electrophoretic progress of a protein in non-denaturing PAGE is determined in part by the protein's conformation. Changes in the conformation due the formation of aggregates (dimers or oligomers) or the binding of ligands to generate more compact (or disperse) structures can be visualized on non-denaturing PAGE.

Table 6.3 *Measurement of enzyme activity in situ following non-denaturing PAGE. Shown are four selected examples of enzyme activities and substrate solutions used to detect in situ activity using chromogenic substrates. For further details and more comprehensive examples, see recommended further reading.*

Enzyme	Substrate solution and method of detection of activity (Total volume 50 ml with an incubation time of 15–20 minutes)
Alcohol dehydrogenase	50 mM Tris pH 8.5 containing, NAD (0.2 mg/ml), 3-(4,5-dimethylthiazol-2-yl)-2,5diphenyltetrazolium bromide (MTT; 0.2 mg/ml), 0.2 mg phenazine methosulphate (PMS; 0.004 mg/ml) and 4 % (v/v) ethanol. Gels are incubated on a shaker in the dark until blue bands appear, at which point they can be fixed in 25% (v/v) ethanol and imaged.
β-Galactosidase	100 mM Na_2HPO_4 pH 7.4, containing 5-bromo-4-chloro-indolyl-β-galactopyranoside (0.2 mg/ml), Nitro blue tetrazolium (NBT; 0.1 mg/ml). Gels are incubated on a shaker in the dark until blue bands appear, at which point they can be fixed in 25% (v/v) ethanol and imaged.
Peroxidase	50 mM Na_2HPO_4 containing NADH (0.11 mg/ml), phenol (0.08 mg/ml), NBT (0.06 mg/ml) and H_2O_2 (0.02 % v/v). Gels are incubated on a shaker until brown bands appear, at which point they can be fixed in 25% (v/v) ethanol and imaged.
L-amino oxidase	65 mM NaH_2PO_4 pH 6.8 containing L-lysine (10 mM), NBT (0.1 mg/ml) and PMS (0.05 mg/ml) on a shaker until blue bands appear, at which point they can be fixed in 25% (v/v) ethanol and imaged.

6.3.2   Cellulose Acetate Electrophoresis

A porous matrix of cellulose acetate is commonly used in the separation of blood plasma proteins in clinical diagnostic tests using short separation times of around 40 minutes. This form of electrophoresis uses a phenomenon known as electro-osmotic flow, caused by the interaction of cations in the buffer with negatively charged groups on the matrix, which can result in the movement of weakly anionic biological macromolecules (e.g. antibodies and other serum proteins) moving towards the cathode.

6.3.3   Western Blotting

In this technique, gels containing proteins separated by either SDS-PAGE, 2D-PAGE or non-denaturing PAGE are transferred electrophoretically from the gel on to a protein binding membrane (e.g. nitrocellulose), which is in contact with it. The resultant Western blot is then probed with antibodies that recognize specific proteins. Such antibodies can be conjugated or 'labelled' with a variety of ligands that facilitate detection under appropriate conditions using imaging instruments. This technique is discussed in more detail in Chapter 9.

6.3.4   Isoelectric Focusing (IEF) and Two-Dimensional Polyacrylamide Gel Electrophoresis (2D-PAGE)

The overall charge on the surface of a protein is pH dependent (see Figure 6.5). At a pH below the pI, the charged amino acid side chains will be more protonated (i.e. more positively charged), whereas at pH values above the pI, the groups will tend to be deprotonated (i.e. more negatively charged).

At a specific point on the pH scale, the *net* charge on the protein surface will be zero. This is called the pI of a protein, where the positive and negative charges on a protein's surface are equalized. It does not mean there is no charge; rather, it reflects the fact that the overall charge is balanced resulting in no net charge overall. At the pH of a protein's pI, the lack of overall charge on the surface of the protein means that

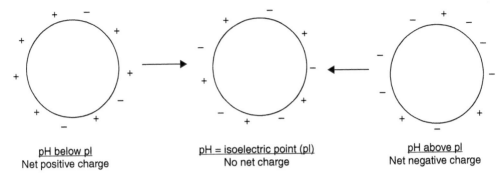

Figure 6.5 *Principles of isoelectric focusing (IEF). During separation by IEF, proteins will have a net negative charge at pH values above their isoelectric point (pI) and a net positive charge at pH values below their pI. This will cause them to migrate towards their pI, as indicated by the arrows, until their net charge is zero.*

the protein will not bind to an oppositely charged particle (e.g. see also Chapter 7, Section 7.11; ion exchange chromatography) and it will not move towards an electrode if subjected to an electric field. This is a property which is useful in the technique of isoelectric focusing (IEF).

In IEF, samples are prepared in a buffer containing a neutral detergent (e.g. CHAPS), reducing agent (e.g. dithiothreitol), a denaturing agent (e.g. 8 M urea) and pH gradient forming molecules called ampholines. They are then applied to a special type of polyacrylamide gel strip containing an immobilized pH gradient (IPG strips) and separated according to their net charge until all polypeptides reach a point where their net charge is zero, which is known as their pI (see Chapter 7, Section 7.11). While this technique can be useful for the separation of proteins (or protein isoforms) of similar size, but with slight differences in net charge, it is more powerful when used in combination with SDS-PAGE as discussed below. This process is shown schematically in Figure 6.6.

In two-dimensional polyacrylamide gel electrophoresis (2D-PAGE), IPG strips containing proteins separated by net charge (pI) are then equilibrated with SDS-containing denaturing sample buffer and separated by differences in size in a second dimension of SDS-PAGE. This approach, which is shown diagrammatically in Figure 6.6, increases the resolution of complex mixtures compared to either IEF or SDS-PAGE alone, and is a major technique used for analyzing the protein profile (proteome) of cells, tissues and sub-cellular fractions (e.g. nuclei, mitochondria, cytosol, and so on). Protein identification would require further analysis on individual polypeptides, using techniques such as Western blotting and mass spectrometry. Images of a cell extract separated by both SDS-PAGE and 2D-PAGE are shown in Figure 6.6.

6.4 Separation of Nucleic Acids by Gel Electrophoresis

At neutral pH, nucleic acids are negatively charged, due to the phosphate groups in the sugar phosphate backbone. As the net charge is similar, due to the regular repeats of phosphate groups, polynucleotide chains vary mainly in length. Therefore, DNA and RNA molecules will migrate in an electric field towards the anode in an electrophoresis gel according to their size. The principles of migration are similar to those for SDS-PAGE except that agarose gels with larger pore sizes are more commonly used for the separation of large DNA or RNA molecules. Thus, the longer DNA sequences will migrate at a slower rate than the shorter DNA sequences.

Manipulation of the agarose concentration determines the effective fragment size separation range, lower concentrations (e.g. 0.3% (w/v)) being better for the resolution of larger DNA size ranges (e.g. 1 000–70 000 base pairs [bp]) and higher concentrations (e.g. 2% (w/v)) being more suitable for smaller DNA size ranges (e.g. 100–5000 bp). Resolution of even smaller DNA molecules may require the use of polyacrylamide (typically 29 : 1 acrylamide: *bis*-acrylamide ratio). For example, a 20% (w/v) polyacrylamide gel will allow

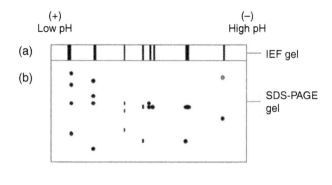

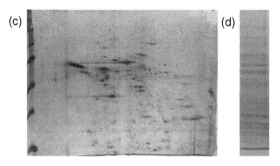

Figure 6.6 *Principles of two-dimensional polyacrylamide gel electrophoresis (2D-PAGE). Shown are schematic representations of a typical isoelectric focusing (IEF) gel, showing how a few polypeptide bands might focus on an IPG strip at their isoelectric points (a). If this IPG strip were to be equilibrated with SDS-PAGE denaturing sample buffer and applied to the top of a SDS-PAGE (b), it can be seen that a number of polypeptide spots of different sizes could be resolved. An example of a real 2D-PAGE separation of a mouse neuronal cell extract, in which the polypeptide profile is revealed by silver staining, is shown in (c). An example of the same extract separated by only SDS-PAGE and then stained with Coomassie blue is shown in (d). It is clear from these examples that many polypeptides can have a similar size or isoelectric point, but they can be resolved from one another if a combined charge and size separation approach is used (i.e. 2D-PAGE).*

the separation of very small DNA fragments between 5 and 100 bp in length, whereas a 3.5% (w/v) polyacrylamide gel would resolve DNA sequences of approximately 100–1000 bp. As polyacrylamide gel preparation has been covered already in this chapter, we will now focus on the preparation and use of agarose gels.

Note: DNA electrophoresis may use toxic chemicals, particularly during the gel staining procedures (e.g. ethidium bromide). You must therefore ensure that you prepare, handle and dispose of toxic reagents in compliance with local health and safety policy. However, alternative dyes such as RedSafe™ have been developed recently which are non-carcinogenic but may cause skin and eye irritation. It has the advantage that it can be visualized not only under UV light but also with visible light, reducing the risk of UV damage when the DNA is to be manipulated in further experiments. It can detect as little as 50 ng of double-stranded DNA in a gel band, which is comparable to the sensitivity of ethidium bromide.

6.4.1 Sample Preparation

DNA or RNA is prepared by specific extraction procedures (see Chapter 5, Section 5.2.2). Before preparing your sample for electrophoresis, it is important to estimate the nucleic acid concentration so that precise amounts of nucleic acid can be loaded onto the gel. This is typically done by estimating the ratio of absorbance 260–280 nm, measured using a spectrophotometer (see Chapter 3, Section 3.8.1).

6.4.2 Preparation and Running of Agarose Gels

Agarose gels tend to be run in a flatbed electrophoresis system in which the gel, which is completely submerged in electrolyte (see Table 6.2), is run horizontally (as opposed to the vertical direction used for the majority of protein separations). A typical gel apparatus is shown schematically in Figure 6.7. Agarose gels tend to be much thicker than polyacrylamide gels as thinner strips of agarose would be very fragile. The agarose is mixed with appropriate buffer components and dissolved by heating, then allowed to form a gel on cooling. As for protein gel electrophoresis, sample application wells are formed by the insertion of a sample well comb before the gel is able to set. This is removed carefully and the wells filled with electrolyte prior to sample application. As for SDS-PAGE, the sample is loaded through the electrolyte aided by the presence of glycerol in the sample buffer (see Table 6.2), which makes the sample sink to the bottom of the well. The current is applied at the recommended level (this varies from one apparatus to another) and the gel is allowed to run until the tracker dye (e.g. bromophenol blue) has moved at least 3/4 of the gel length.

6.4.3 Staining of Gels to Detect Nucleic Acids

Gels are carefully removed from the gel tank with gloved hands and immersed for approximately 5 minutes in a solution containing the fluorescent nucleic acid binding dye ethidium bromide or a suitable (safer) alternative, such as SYBR®Safe, GelRed™, GelGreen™ or EvaGreen®, which have been shown to be less carcinogenic than ethidium bromide. These dyes intercalate between paired nucleotide of sections of double helix in DNA. The gel is then carefully transferred to a UV trans-illuminator which may be part of an

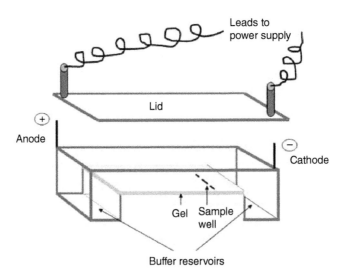

Figure 6.7 *Schematic diagram of a typical apparatus for agarose gel electrophoresis. A typical apparatus comprises two buffer reservoirs either side of a raised platform onto which the agarose gel is polymerized. The samples are applied into the sample wells shown and electrophoresed in a horizontal direction. Nucleic acids migrate towards the anode due to their intrinsic negative charge due to the phosphate groups. The buffer tank is filled until the gel is submerged in the electrolyte.*

imaging instrument which encloses it in a sealed unit. It is important to wear protective glasses, particularly if a sealed imager is not available, as there is a risk that your eyes may be overexposed to UV. Bands can be observed by a special camera that detects UV light and a digital or photographic image recorded. DNA or RNA molecules are visible as fluorescent bands that migrate according to size differences. In a similar fashion to SDS-PAGE, running agarose gels with a series of standards (polynucleotides of a size range that covers the range of resolution of a particular gel) in parallel allows the estimation of molecular weight from a standard curve like that used in SDS-PAGE. However, remember that in the case of nucleic acids, molecular weight is estimated as the number of base pairs (bp). A more precise evaluation of molecular weight would require sequence analysis to take into account the exact base composition.

6.5 Applications of Gel Electrophoresis of Nucleic Acids

6.5.1 DNA Fingerprinting and PCR

This involves the electrophoretic separation of DNA fragments produced either by reverse transcriptase polymerase chain reaction (RT-PCR) or restriction endonuclease-mediated digestion of DNA. In the case of RT-PCR, the amplified product (amplicon) is often eluted from the gel and sequenced to confirm that the appropriate gene sequence has been amplified. When the digested DNA is from whole DNA, it can be used to determine the DNA fingerprint of an individual or species or to identify polymorphisms in genes (i.e. changes in the sequence of DNA that cause changes in endonuclease cleavage resulting in a different DNA fragmentation pattern). An image of a typical stained agarose gel is shown in Figure 6.8.

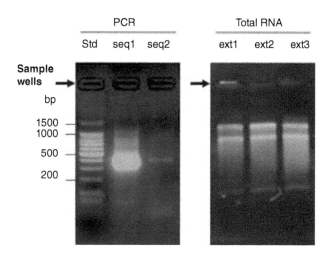

Figure 6.8 *Examples of agarose electrophoresis gels stained with ethidium bromide. The agarose gel on the left hand side shows the molecular weight standards (Std: a ladder of polynucleotides of different sizes) and the PCR amplification products of two different mouse gene sequences of similar length (seq1 and seq2). Both sequences have a size of approximately 500 bp. The gel on the right shows the separation of total RNA from three different plant cell extracts. A consistent pattern can be seen in each sample, the two strongest bands representing the ribosomal RNA subunits. The streaky 'tail' below the smaller (lower) of the two bands reflects the many different sizes of messenger RNA sequences that would be expected to be present in a cell or tissue extract. The two gels are not directly comparable as a much lower agarose concentration was used for the gel on the right.*

6.5.2 Analysis of RNA

Total cellular RNA, ribosomal RNA and purified messenger RNA molecules can be separated using agarose gel electrophoresis. The quality of total RNA preparations can be assessed by the appearance of the larger stronger staining bands corresponding to ribosomal RNA subunits (see Figure 6.8). However, it is important to note that under conditions used for double-stranded DNA separation, RNA molecules may exhibit secondary structure that causes abnormalities in their electrophoretic migration, making accurate molecular weight determination more difficult. Secondary structure can be eliminated by performing denaturing gel electrophoresis for which the samples are pretreated by heating in the presence of formamide or glyoxal and electrophoresed using buffers containing formaldehyde.

6.5.3 Pulsed Field Gel Electrophoresis (PFGE)

This technique allows separation of DNA sequences of up to 12 Mbp in length and allows researchers to gain structural information about very large regions of genomic DNA (i.e. DNA containing both coding and non-coding sequences). These regions of DNA are produced by digesting genomic DNA with restriction endonucleases that cut at very rare cleavage sites containing eight bp, resulting in the production of much larger fragments of DNA than would be produced by more commonly used restriction enzymes, which cleave at sites between four and six bp in length. Pulsed field gel electrophoresis is different from the more conventional method described earlier in that it uses at least two alternating electric fields. These fields are arranged at different angles in such a way that results in very effective size-based separations, due to the fact that larger DNA molecules take longer to reorientate to the field direction change than do smaller ones.

6.5.4 Identification of Specific Sequences Using Hybridization Techniques

After conventional agarose gel electrophoresis, separated nucleic acids can be transferred onto a membrane filter (usually nylon or nitrocellulose) to produce a replica of the gel separation pattern. To facilitate transfer to the membrane, the gel is first of all soaked in HCl (which randomly cleaves DNA into smaller pieces) then with alkali (which denatures the double-stranded DNA to single-stranded polynucleotide chains) and then neutralized. Alternatively, DNA can be fragmented at specific points along its polynucleotide sequence using enzymes termed restriction endonucleases, and the resultant fragments separated by agarose gel electrophoresis.

Notes on Restriction Endonucleases

- These are enzymes that can cleave double-stranded DNA along both strands. 'Restriction enzymes' recognize specific sequences at which they are able to cleave the DNA without damaging the bases.
- This approach has been a cornerstone in recombinant gene technology, as the resultant DNA pieces can be joined up (ligated) with other DNA segments cut with the same restriction enzyme, using DNA ligases.
- Examples of restriction enzymes and their cleavage sites on double-stranded DNA are shown below.

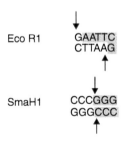

- The shaded and unshaded areas represent the ends of the two cleavage products.
- Note that the ends can be uneven ('sticky ends') or straight ('blunt ends').
- As the occurrence of a specific cleavage site is unique to each gene, this approach is also applied in DNA fingerprinting in forensic science and in the analysis of restriction fragment polymorphisms which can help to identify gene mutations.

The gel is then placed in close contact with the membrane filter and several layers of filter papers which soak up the buffer and nucleic acids in the gel through the membrane by capillary action, leaving the nucleic acid sequences bound to the membrane as they come into contact with it. This process can be speeded up by using positive pressure, a vacuum or an electric field.

The blot is then fixed by heating in a drying oven before specific stretches of DNA or RNA are identified by incubation of the blot with complementary nucleotide sequences with which they will hybridize (bind). These probes are either radiolabelled during synthesis (detected by autoradiographic exposure onto X-ray film) or have a non-radioactive label (e.g. digoxigenin) that can be detected by a chemical reaction. For example, blots can be probed with horseradish peroxidase-conjugated anti-digoxigenin antibodies. This could be followed by the addition of an appropriate substrate to detect conjugated enzyme activity either by the formation of an insoluble coloured reaction product or by the release of chemiluminescence.

This kind of approach can be used for probing DNA (Southern blotting) or RNA (Northern blotting). The combined use of gel electrophoresis and blotting gives information about the size of a specific sequence. However, if only detection is required, an alternative to the above is to use dot or slot blotting, in which the nucleic acid extracts are pipetted directly to the membrane through round (dot) or slot-shaped template holes, then probed as for the Northern and Southern blots.

6.5.5 DNA Sequencing

The sequence of DNA can be determined by performing a procedure known as the Sanger reaction on a given gene or part of a gene sequence. This involves the use of single-stranded DNA as a template for DNA synthesis in the presence of low levels of radiolabelled or fluorescently labelled dideoxynucleotides which, as they are unable to form the phosphodiester bonds in the DNA backbone, act as a chain terminator. If separate reactions are performed with all four dideoxynucleotides (ddA, ddT, ddG and ddC) at low concentrations, some polynucleotide chains will lengthen as normal, whereas others will terminate at each stage that the normal deoxynucleotide is present, leading to a series of chain lengths for each dideoxynucleotide corresponding to each position where the corresponding deoxynucleotide is found. Such chains are separated on denaturing polyacrylamide gels and the bands detected by autoradiography (radioisotopes) or by laser scanning (fluorescence). The sequence of the template is then deduced from the position of the bands (which represent chains terminated at each point that the corresponding deoxynucleotide should be present) in each of the four sample wells from top (3') to bottom (5') ends of the gel. Remember that the sequence determined initially is the complementary sequence to the original template, thus the true sequence is determined by its complementary base pairing (e.g. A-T and G-C). A schematic representation of a sequencing gel is shown in Figure 6.9, together with an indication of how the sequence is determined from the band profile on the gel.

6.6 Summary

- Gel electrophoresis is a means of separating proteins and nucleic acids in an electric field through a three-dimensional gel matrix.
- Separation is dependent on differences in mass, shape and charge of the separated molecules and on the resistance experienced during electrophoretic migration.
- Polyacrylamide gels are used mainly for protein separations but also for small polynucleotides, whereas agarose gels are the norm for larger nucleic acid sequences.

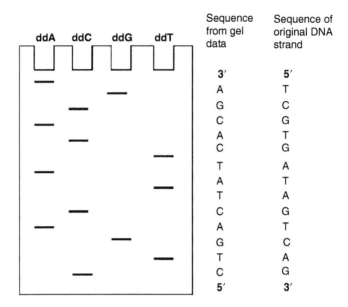

Figure 6.9 *Schematic representation of a DNA sequencing gel. Shown is a schematic representation of a typical gel pattern detected by autoradiography of a sequencing gel. Above each sample well is an indication of which dideoxy nucleotide was added in the reaction described in Section 6.5.5. Each band represents a chain length at which the polynucleotide synthesis came to a halt due to the addition of a dideoxynucleotide instead of the deoxynucleotide that would be added during normal DNA synthesis. The positions of the bands in the four different banding patterns can then be used to determine the position of each consecutive nucleotide in the polynucleotide sequence. The sequence of the original template is then derived from the known pattern of base pairing, as shown.*

- Electrophoresis has major applications in the characterization of isolated proteins and nucleic acids (e.g. determination of molecular weights and purity of purified proteins, nucleic acids, PCR products and in DNA sequencing).
- It can be combined with blotting techniques that allow probing of protein or nucleic acid samples transferred onto membrane filters for specific components, using antibody or polynucleotide probes.

7
CHROMATOGRAPHY

7.1 Introduction

Chromatography was first developed by the Russian botanist Mikhail Tswett in the early 1900s when he produced a colour abundant separation of plant pigments. The term chromatography can be translated literally from the Greek words *chroma* and *graphein* which means to 'write with colours'. The technique remained peripheral to mainstream analytical techniques until the 1930s, when paper and thin-layer chromatography (TLC) became popular. In the following decades, many novel chromatographic techniques were developed, making chromatography the most widely used analytical technique in the biosciences.

The diversity of chromatographic methods means that chromatography can be used to purify any soluble (or volatile) compound if the correct adsorbent material, mobile phase and operating methods are employed. Chromatography can separate complex mixtures with impressive resolution; components such as proteins that may vary by only a single amino acid can be separated by chromatography. These components are also immediately available for identification and quantification. If an important component has been identified, the whole chromatographic procedure can be 'scaled up' from a small-scale analytical run to a preparative run. Also, the conditions employed for chromatography are typically not severe (gas/liquid chromatography (GLC) is an exception), which allows for the analysis of sensitive components.

7.2 The Theory of Chromatography

All compounds have different structures, and the properties of these different structures will influence the interaction a compound has with its immediate environment. If a compound is placed into an environment containing two immiscible phases, the properties of the compound's structure will determine which of the phases (or both phases) the compound will preferentially interact with. A distribution equilibrium (K_D; also called the partition coefficient) will be established, which is a measure of a compound's capability to interact with the two immiscible phases. It is a ratio of the concentration of a compound in one phase divided by its concentration in the other phase.

Compounds can be separated from each other in a simple laboratory apparatus called a separating funnel, into which two immiscible liquid phases have been placed. The separating funnel is shaken to evenly distribute the compound between the two phases and to allow an equilibrium to be established. When the phases have separated in the funnel, they can be isolated into separate collecting vessels. The concentration of the compound in each of the phases can be measured (see Figure 7.1) and the partition coefficient (K_D) determined.

Separating funnels produce one partition event, which may be insufficient to gather most of the compound of interest into one phase. To improve the separation and the yield of the compound of interest, the experiment can be repeated with fresh liquid. The use of separating funnels is a time-consuming process; however, chromatography columns can reproduce these events many times in quick succession. In chromatography, the two immiscible phases are the *mobile phase* (solvent) which carries the sample over and sometimes through the *stationary* phase.[1] These phases may be combinations of liquid and liquid, gas and liquid or liquid and solids. Mixtures of compounds introduced into this flowing system partition between the two immiscible phases, and they will be separated if they have different partition coefficients. Sample components that preferentially partition into the stationary phase will spend a greater amount of time on the column (prolonged elution time on the chromatogram). The components favouring the stationary phase are

Basic Bioscience Laboratory Techniques: A Pocket Guide, Second Edition. Philip L.R. Bonner and Alan J. Hargreaves.
© 2022 John Wiley & Sons Ltd. Published 2022 by John Wiley & Sons Ltd.

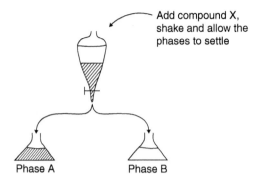

Add compound X, shake and allow the phases to settle

Phase A   Phase B

The amount of compound X in each phase can be measured and the partition coefficient (K_D) determined

$$(K_D) = \frac{\text{the concentration of compound X in phase A}}{\text{the concentration of compound X in phase B}}$$

Figure 7.1 *Separating funnels can be used to determine a partition coefficient (K_D).*

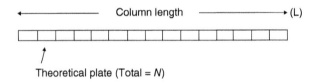

Column length ⟶ (L)

Theoretical plate (Total = N)

Figure 7.2 *Theoretical plates in chromatography columns.*

separated from other components that stay predominantly in the mobile phase because the mobile phase components will pass through the column quickly (reduced elution time on the chromatogram or data station readout). The detector at the end of the column will respond proportionally to the concentration of the compound in solution, marking the presence of the compound as a peak with a characteristic retention time (or volume). Every compound applied to a chromatography column will emerge as a 'normal' (Gaussian) distribution (see Figure 7.3 and Chapter 4, Section 4.3.1).

7.2.1 The Partition Coefficient (K_D) in Chromatography

$$K_D = \frac{\text{the molar concentration of a compound in the stationary phase}}{\text{the molar concentration of a compound in the mobile phase}}$$

The length of column where an individual equilibrium between the sample in the mobile phase and the stationary phase takes place is called a theoretical plate (there will be a total of N theoretical plates in a column; see Figure 7.2 and Worked Example 7.1). The length of the column containing a theoretical plate is called the plate height (H; unit of distance, e.g. mm). The more times a compound can establish equilibrium on the column, the more likely it will be resolved from other components. Therefore, a good chromatography column will have many theoretical plates (see Worked Example 7.1) and a small value (e.g. mm) for the height equivalent to theoretical plate value (HETP).

Although theoretical plates are a useful numerical measure of a column's efficiency, it is important to remember that they do not actually exist. The **plate** theory of chromatographic separation assumes an

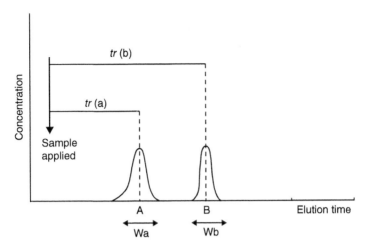

Figure 7.3 *The elution profiles of compounds A and B on a chromatography column. The time the compounds have been retained on the column (tr) and the peak width (W) can be measured.*

almost instant interaction between a compound in the mobile phase and the stationary phase. The plate theory also assumes that the movement of compounds down the column is by the transfer from one theoretical plate to the next theoretical plate and this process is repeated until the end of the column is reached.

However, compounds applied to a column take time to associate and interact with the stationary phase. The **rate theory** of chromatography considers other factors that contribute to the final elution profile of a compound in chromatography experiment.

- The width of a chromatographic peak (see Figure 7.3) is linked to the flow rate of the mobile phase (u).
- The molecules of any compound applied to a column will take different random paths through the stationary phase which contributes to peak broadening (eddy diffusion) (A).
- The concentration of any compound is greatest at the centre of the peak, and this concentrated material will tend to diffuse out towards the edges, which will also contribute to peak broadening (longitudinal diffusion) (B).
- When a compound has a strong interaction with the stationary phase and the flow rate is too high, any compound remaining in the mobile phase will be swept ahead of the compound interacting with the stationary phase, resulting in peak broadening (resistance to mass transfer) (C).

The van Deemter equation relates the factors above to the height equivalent to a theoretical plate (HETP) of a column and stationary phase.

$$\text{HETP} = A + \frac{B}{u} + Cu$$

Worked Example 7.1 Resolution, Peak Width and Theoretical Plate Number

Two compounds were applied to a 15 cm column. The retention time for compound A was 5 minutes, and the base line peak width was 12 seconds. The retention time for compound B was 8 minutes with a peak width of 10 seconds.

Chromatographic columns are judged by their ability to provide baseline separations of different components in sharp peaks (see Figure 7.3). Two important parameters are resolution (R) and the peak width (W).

$$R = \frac{\left[2\left(tr(b) - tr(a)\right)\right]}{\left[W(a) + W(b)\right]}$$

$$R = \frac{\left[2(480 - 300)\right]}{12 + 10}$$

$$= 16.36 \text{ seconds}$$

- A number >1.5 usually means baseline separation. Clearly, these peaks (see Figure 7.3) are well-resolved, and a visual check of a chromatogram is always recommended.
- The efficiency of the separation by a chromatography column of individual components is measured in theoretical plate number (N).

$$N = 16(tr / W)^2$$

For compound A,

$$N = 16(300 / 12)^2$$
$$= 10000$$

For compound B,

$$N = 16(480 / 10)^2$$
$$= 36864$$

Chromatography columns act as if they have different plate numbers for different compounds.

- When two peaks elute closer together, the peak width at half peak height ($W_{0.5h}$) can be used to calculate the theoretical plate number.

$$N = 5.54(tr / W_{0.5h})^2$$

The height equivalent to a theoretical plate (HETP) allows comparisons to be made between columns of different lengths (L) or the same column run under different conditions.

$$HETP = L / N$$

Column efficiency should improve with an increase in N or a decrease in the HETP.

7.3 Factors to Consider in Chromatography

- The manufacturers of chromatography columns will indicate the optimal flow direction of the column. Attach the column to the HPLC in the recommended flow direction.
- All compounds elute from a chromatographic column in a manner closely approximating to a 'normal' (Gaussian) distribution, and they will have a characteristic retention factor (time, volume or distance moved) (see Figure 7.3).
- Identification of compounds in a sample can be achieved by comparing the retention time of a known standard to the retention time of components in the sample (this does assume some prior knowledge of

the type of compounds to be separated). However, a similar retention time to a known standard is not a categorical proof that the sample separated is, indeed, the same as the standard used. Quantification can be achieved after chromatographic separation by applying the Beer Lambert law (see Chapter 3, Section 3.6) using the UV/vis absorbance set at the required wavelength as the detection procedure (a diode array spectrophotometer can measure wavelengths 200–800 nm throughout the entire run. This generates a lot of data).

- Incrementally increasing the amount of a standard compound injected onto the column will produce corresponding increases in the peak area of the standard compound. A plot of peak area (y-axis) against the amount of the standard compound will produce a calibration plot which can then be used to quantify the amount of a compound in a sample.

- The addition of a known amount ('spike') of the standard to the sample before chromatography will help to identify and quantify the amount of the compound of interest in the sample.

- Other techniques (e.g. mass spectrometry and/or NMR) will be needed to validate an unknown compound in a sample. Many chromatography systems have dual detectors e.g. UV/vis and mass spectrometer.

- Most modern chromatographic systems have data management devices (computers or integrators) linked to the detector (see Figure 7.7), which can be used to measure the peak areas and produce calibration graphs.

- However, if the chromatography system lacks a computer but is linked to a chart recorder, the peaks from the standard and the samples can be cut from the chart recorder paper with scissors and weighed to construct a calibration graph.

- Applying crude samples to a column will not only complicate the analysis but can also lead to some components permanently bonding to the HPLC resin in the column; this will reduce the column's capacity (the amount of sample which can be applied to the column without altering the resolution). Solid-phase extraction (SPE) is a popular procedure to remove 'column damaging' compounds and to enrich the components of interest. Cartridges of various dimensions can be purchased and used with a syringe to quickly clean up the sample mixture. If a compound is known to elute from a reversed phase (RP) C_{18} (ODS, octadecylsilane) analytical column at 50% (v/v)[2] methanol, a C_{18} SPE cartridge can be used in the sample preparation (see Worked Example 7.2 and Section 7.5.1).

- To overcome any variation in sample preparation and to counter the small changes in the amount of sample applied to the column, a known amount of a compound can be added to the extract. This internal standard should be chemically related to the compounds of interest, but it should also ideally elute in a different part of the chromatogram. Variation in the extraction or injection process can then be compensated for by reference to the constant area of the internal standard.

- Use the highest grade of solvents and water available to make up chromatographic reagents.

- Always filter the reagents through a 0.2 μm filter to remove particulate matter (and bacteria) which could clog up the pores in the column.

- When there is a switch in solvent from 100% solvent to 5% solvent plus 95% water, there will be an increase in back pressure. To protect the column from adverse pressure surges, set a maximum pressure limit (below the maximum allowed) on the chromatographic system or decrease the flow rate to lower the back pressure.

- Having taken great care in the preparation of your samples and in the filtration of all the solvents, after prolonged use, an HPLC column will start to show an increase in back pressure. This may be due to particles smaller than 2 μm in diameter present in the sample or the solvents or caused by small particles of the HPLC gaskets being released into the solvent over time becoming trapped in the spaces surrounding the HPLC column resins.

- When an increase in the back pressure is detected on an HPLC system, the first item to check is the guard column. This can be replaced or cleaned before refitting into the HPLC system.

- Cleaning and regenerating C_{18} RP columns that have been used frequently may be achieved by first reversing the flow of the solvent through the column, without the solvent passing through the detector. Allow the solvent to flow in the opposite direction to the recommended flow direction at a slow flow rate (e.g. 0.1 ml min^{-1}) in 100% methanol for 60 minutes.

- Return the column to the correct flow direction and monitor the back pressure.

- If the back pressure does not return to an acceptable pressure rating, the column can be washed (without the flow passing through the detector) with the following sequential protocol: 10 column volumes of methanol, 10 column volumes of acetonitrile, 10 column volumes of isopropanol, 10 column volumes of acetonitrile, 10 column volumes of methanol and finally 10 column volumes of 10% (v/v) methanol or the starting buffer for the procedure in use. Monitor the back pressure at all stages and reduce the flow rate if the back pressure starts to rise.

- To help the column cleaning, increase the temperature of the column using a column heater or immerse the column in a water bath set to 60 °C
- At the end of the day or session, a 'saw tooth' wash can be routinely used to help maintain the HPLC columns life span. This 'saw tooth' wash is a gradient from the start conditions, e.g. 5% solvent (B) to the end condition, e.g. 100% solvent (B) over 30–60 minutes at a slow flow rate (e.g. 0.1 ml min⁻¹). This is followed by a gradient from 100% solvent (B) to 5% solvent (B) over the same period. The process can be repeated many times overnight at a slow flow rate e.g. 0.1 ml min⁻¹. The column starting conditions can be timed to be ready for injecting the next sample the following day (see Figure 7.4d).

7.4   The Methods Used to Elute Samples in Chromatography

- **Isocratic elution:** The % eluent of the mobile phase remains constant for the elution period (e.g. 2–5 column volumes). At the end of the isocratic wash period, the % eluent of the mobile phase may be altered for another isocratic wash (see Figure 7.4a). This is a useful technique to employ when the sample components elute close together in a gradient elution.
- **Gradient elution:** The % eluent of the mobile phase is gradually altered to increase or decrease one of its components (see Figure 7.4b and 7.4c).
- **Displacement elution:** It involves washing the column with a molecule that has a higher affinity for the ligand on the stationary phase than the molecule bound to the resin. Displacement elution can be used in ion exchange chromatography (see Section 7.8) and affinity chromatography (see Section 7.10). The eluted component migrates down the column as a front rather than the normal distribution seen in gradient or isocratic elution.

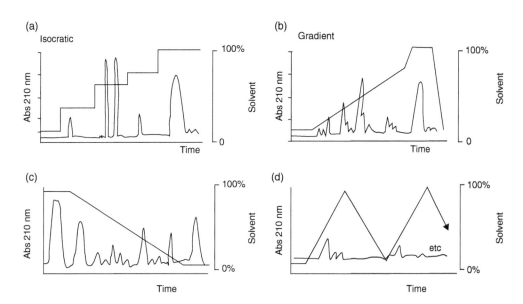

Figure 7.4 *Common solvent profiles that are used to elute samples from uHPLC or HPLC chromatography columns. (a) A stepped isocratic solvent elution; (b) a positive linear solvent gradient; (c) a negative linear solvent gradient used in HILIC and (d) a 'saw tooth' solvent gradient used to wash a chromatography column.*

Worked Example 7.2 A Typical Procedure for the Preparation and Use of a C_{18} SPE Column in Sample Preparation.

- If a component in a sample has been characterized to elute at a particular %methanol, e.g. (50%), the use of the following SPE sample preparation protocol can be of value to protect an analytical HPLC column. This method can be easily adjusted to accommodate the preparation of most components.
- The SPE cartridge is first washed in methanol before being equilibrated in 10% (v/v) methanol.
- Methanol is added to the sample to a final concentration of 10% (v/v) before application to the SPE cartridge.
- The SPE cartridge is then washed with 10% (v/v) methanol to elute components that do not bind to the resin.
- The sample can then be eluted in 70% (v/v) methanol and dried (see Section 7.5.4) or diluted before being applied to the analytical C_{18} HPLC column.
- Any damaging components will remain bound to the SPE cartridge rather than binding irreversibly to the expensive analytical C_{18} HPLC column.
- The SPE cartridge can then be disposed of in an appropriate fashion.
- A syringe is fine for individual samples, but for processing large numbers of samples, the SPE cartridges can be loaded into proprietary SPE racks, which use vacuum pumps to quickly load and elute the samples.

7.5 A Selection of Methods that Can be Used to Exchange Buffers and Concentrate Samples Prior to Chromatography

7.5.1 Dialysis

The procedure of dialysis is used extensively to support people with clinical urology problems. When a person has acute kidney problems, they may be referred for regular blood dialysis. The loss of kidney function results in the build-up of detrimental urea and creatine in the blood; these chemicals are routinely filtered out of the body by the kidneys of healthy individuals. The dialysis procedure flows the patient's blood through a membrane which has a molecular weight cut-off (MWCO) of approximately Mr 12 000. This means that most proteins and the cells (Mr > 12 000) in the blood of a patient with kidney problems are retained in the blood and returned to the patient's body (see Figure 7.4a). Molecules with a relative molecular mass Mr < 12 000 (e.g. urea, creatine, metabolites, metal ions and peptides) pass through the semipermeable dialysis membrane and are removed from the patient's blood. The cells in the glomeruli of the kidney are selectively permeable and are able to retain beneficial small metabolites. The dialysis process is monitored throughout and vital metabolites may be replaced before a problem developes.

In the laboratory, when an extract contains small molecules that need to be removed, e.g. prior to an enzyme assay, dialysis can be used. Dialysis tubing is typically constructed from thin cellulose or cellulose ester films (dialysis membranes can also be constructed from polyether sulphone and collagen used in specific applications). When hydrated, the dialysis membrane can be fashioned into a bag (see Figure 7.5b) into which the extract can be placed. When the dialysis bag is sealed, it can be placed in a large volume of water or an alternative buffer to remove the contaminating small molecules. The process is thermodynamically driven because molecules at a high concentration (relatively low entropy S) will disperse to an area of low concentration where they have increased entropy (S). Over time, an equilibrium will be established such that the concentration of small molecules will be the same inside to outside the dialysis bag.

At equilibrium, there will be no net movement of molecules from inside to outside of the dialysis bag. To avoid the point of equilibrium, the volume of liquid outside of the dialysis bag can be replaced regularly. A typical regime is to have the dialysis bag constantly stirring (to help establish the equilibrium at a faster pace) and to replace the liquid outside of the dialysis bag on a regular basis (e.g. every 2.0 hours). Over a period of time, the concentration of contaminating small molecules will be reduced to an acceptable level. Membranes are available (e.g. Thermo Fisher, UK) with different MWCOs, which may be appropriate in certain circumstances. However, although the process of dialysis is effective, it remains time-consuming.

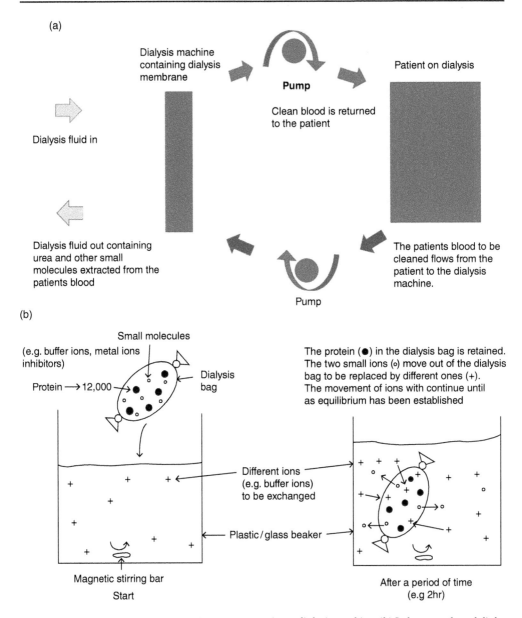

Figure 7.5 *(a) A patient receiving routine treatment using a dialysis machine. (b) Laboratory-based dialysis using regenerated cellulose dialysis membrane. This can be used to exchange unwanted ions and molecules with relatively small molecular masses (Mr < 12 000) in the laboratory. (a) Taken from: https:// www.guysandstthomas.nhs.uk/news-and-events/2012-news/20121022-borough-dialysis-centre.aspx.*

7.5.2   Size Exclusion Chromatography

The use of size exclusion chromatography (SEC) (see Sections 7.5.4 and 7.10) to remove small molecular weight contaminants in a biological extract is a relatively inexpensive and rapid procedure. The process may be used to replace the buffer in a protein solution or to remove small inhibitory metabolites from a protein extract prior to an enzyme assay. The use of chromatography resins with a low MWCO, e.g. Sephadex G-25 (MWCO approximately 1500) and G-50 (MWCO approximately 30 000) (GE Healthcare, UK) or Biogel P6 cartridges (MWCO approximately 6000; BioRad UK), ensures that protein is excluded from the porous network of the size exclusion resins eluting rapidly in the void volume (Vo). The metabolites, ions or buffer components in the extract will have access to the porous network of the resins structure and will elute in the total volume (Vt) of the SEC column away from the volume containing the protein. When the volume of SEC resin is five times larger than the volume of sample to be treated, the process of buffer exchange (metabolite or ion removal) will be complete. The process of SEC using analytical resins (see Section 7.10) will also provide buffer exchange, metabolite or ion removal, but the process requires a longer time period. It should also be remembered that both dialysis and SEC will dilute the sample by increasing the sample volume

7.5.3 Ultrafiltration

Filtration using filter paper is a versatile technique, used in a laboratory to remove large particulate matter or collect precipitated crystals in a liquid. The process relies on gravity to draw the liquid through the pores in the filter. The mass of the particles in the liquid prevents their movement through the pores in the filter, and the particles are retained on the surface of the filter paper. The pores of laboratory filter paper (approximately 20–25 μm diameter) are too large to retain bacteria or mitochondria (approximately 0.5 μm diameter) or proteins (approximately 2–10 nm diameter). However, membranes have been constructed from cellulose acetate, polysulphone, polypropylene, polylactic acid and ceramic material with pore sizes in the range of 5–100 nm diameter. This allows the retention of bacteria, cellular organelles and proteins with different mass ranges. The force of gravity is insufficient to drive the liquid through these membranes with small pores; additional pressure is required to aid the process of filtration which is now described as ultrafiltration. The process of ultrafiltration depends upon the size and shape of the particle and less on the charge of the particle.

Ultrafiltration is utilized in the glomeruli of the kidneys to filter out small molecules present in the blood, e.g. urea, and retain larger molecular structures, e.g. blood cells and proteins (see above Section 7.5.1). The pressure for ultrafiltration in the human body is provided by the heart, which pumps blood around the body at 2.3 psi (120 mm Hg). In the laboratory, the additional pressure for ultrafiltration can be provided by nitrogen gas at 55–70 psi using stirred cells (see Figure 7.6a) or by relative centrifugal force (RCF: sees Chapter 5, Section 5.5) using centrifugal concentration devices. These centrifugal concentration devices (Millipore, Sartorius, Viva products and Pall, UK) have filters with different MWCOs and can accommodate different sample volumes of samples from 1 to 50 ml. The use of a centrifuge allows the processing of many samples at the same time.

The sample chamber of the centrifugal concentrating devices contains a MWCO filter which is positioned facing away from the rotating axis of the centrifuge (see Figure 7.6.b). When the centrifugation is initiated, the increase in the force of gravity drives the liquid towards the surface of the filter. This forces molecules smaller than the MWCO of the filter membrane through the membrane and retains molecules with larger relative molecular mass (Mr). As the exact time required to concentrate a biological sample will vary, the experimental conditions would need to be optimized for the extract of interest.

7.5.4 Centrifugal Evaporating Machines

Biological laboratory centrifugal concentrator machines (Genevac, SciQuip, Eppendorf UK.) are appliances designed to remove liquid (organic solvent or water) from samples to concentrate them to a smaller volume or to complete dryness. These machines can accommodate different sample tubes, from polymerase chain reaction tubes (0.2 ml) to 50.0 ml centrifuge tubes. When the samples are loaded into the sample chamber, the centrifuge starts turning the samples at high speed, encouraging evaporation of the liquid at the sample's surface. The chamber can also be heated to promote evaporation of the liquid (heating of protein samples must be treated with caution). At pre-set intervals, a volume of air within the chamber is removed by a vacuum pump and passed over a cold condensing chamber (dry ice or liquid nitrogen) to

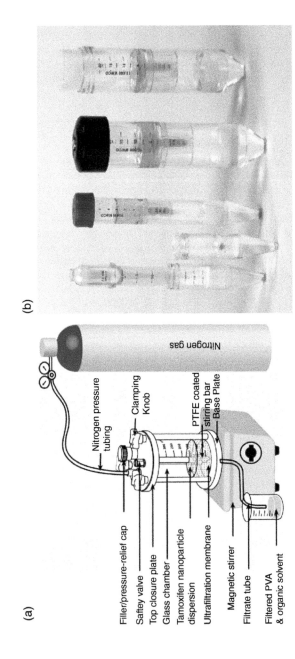

Figure 7.6 (a) An ultrafiltration stirred cell which could be used in a laboratory. Taken from: https://www.semanticscholar.org/ paper/Purification-of-Drug-Loaded-PLGA-Nanoparticles-by-Paswan-Saini/8640a20f10e3a5c252828cf5e73e2d0f4d61bb55/ figure/4. (b) Centrifugal concentrators which can accommodate different volumes of sample. Taken from: https://www. sigmaaldrich.com/GB/en/product/sigma/z614467.

Figure 7.7 *A Genevac 'Mivac' centrifugal evaporating set-up. Source: Genevac Inc.*

remove the solvent/water (see Figure 7.7). All liquids move into the gas phase rapidly at a lower pressure, with solvents drying down more quickly than water. Refer to the manufacturers of the centrifugal concentrator machine for their recommendations on the temperature and running time for different solvents.

7.5.5 Lyophilization (Freeze-Drying)

Water can be removed from biological samples by transporting water molecules from a frozen state to the gaseous state by a process called sublimation. The sublimation of ice to the gas phase occurs at all temperatures below freezing, but sublimation is accelerated by reducing the air pressure (e.g. 0.001 atmospheres pressure: atm) when the sample is at a low temperature (typically <−80 °C). Lyophilization will dehydrate biological tissue and solutions of biological molecules. This is best achieved by rapidly freezing the sample in liquid nitrogen (or dry ice) to prevent the harmful effects that may occur in the slow freezing of a sample when they are placed in a −20 °C freezer, e.g. ice crystals may form in the slow freezing of samples which may damage the integrity of the sample. After rapid freezing, the sample is placed into the sample chamber (see Figure 7.8) before the vacuum pump is switched on. The air in the sample chamber (containing water molecules from the sample) is constantly removed by action of the vacuum pump. The water molecules in the air that are removed from the sample chamber are trapped (refrozen) in a freezing chamber before the air enters the working parts of the vacuum pump. The air from the chamber is finally discarded from the vacuum pump and is voided as exhaust (Local health and safety issues should be adhered to see below Section 7.6). The lyophilization process can be difficult to judge, but one rule of thumb is that as the water leaves the sample in the chamber, the temperature of the sample will rise. When the vacuum pump is switched off and the chamber opened if the sample feels cold, the lyophilization process is incomplete. The sample must be returned to the chamber, the vacuum pump must be switched on and the process be left for a longer period. Many metabolites and proteins are lyophilized (freeze-dried) to help prolong their shelf life.

7.6 Vacuum Pumps

Vacuum pumps can be divided into two types: rotary vane pumps that require oil and oil-free diaphragm pumps. The solvents in some samples must be removed before contacting the rotary vanes of the oil vacuum pumps. The oil within the pumps can be polluted with the solvents, reducing the efficiency of the rotary vane pumps. They require a cold trap to help remove the solvents from the evacuated air prior to its contact with the oil. Even with the best of care, oil-based vacuum pumps require regular oil changes and an oil mist condenser in the vacuum outlet to prevent oil mist in the vacuum pump exhaust from entering the atmosphere. The oil-free diaphragm vacuum pumps are initially more expensive, but the maintenance costs are lower.

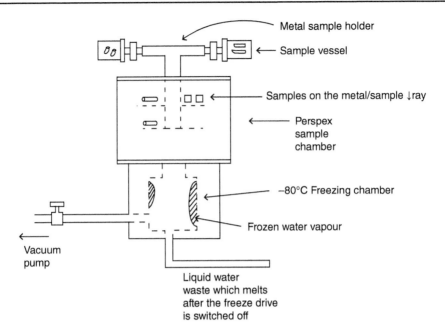

Figure 7.8 *A schematic drawing of a freeze dyer.*

7.7   Different Types of Chromatography and What Properties Can be Used to Separate Molecules

Chromatography is a method that relies on the *differences* in partitioning behaviour (K_D) between a flowing mobile phase and a stationary phase to separate the components in a mixture (see Figure 7.5). Molecules can be separated from one another if there are differences in one or more of the following properties:

- Molecular size and shape
- Surface charge (the presence of positive and negative charges)
- Polarity (a measure of how hydrophobic or hydrophilic a compound is)
- Volatility (how easy does it become a gas).

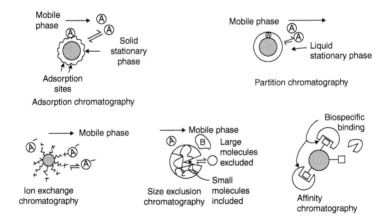

Figure 7.9 *Different types of chromatography.*

The differences in these properties can be exploited by chromatographic methods to separate components in a complex mixture (as outlined below).

7.7.1 Adsorption Chromatography

Adsorption chromatography (see Section 7.8) allows for sample compounds (analytes) to interact with adsorption sites on the stationary phase. In adsorption, the sample interacts only with the surface of the stationary phase (mainly by van der Waals interactions). The interaction of the analytes at the surface of the stationary phase allows an equilibrium to be established with the mobile phase. This can be a solid/liquid interaction, for example, TLC or a solid/gas interaction, as in GLC.

7.7.2 Partition Chromatography

Partition chromatography (see Figure 7.9) allows for a sample compound to dissolve (partition) and establish an equilibrium between the liquid bonded to the stationary phase and the mobile phase. This is usually a liquid/liquid interaction, for example, normal, HILIC and RP chromatography.

7.7.3 Ion Exchange Chromatography

Ion exchange chromatography (see Section 7.11) allows for samples with a positive or negative charge to interact with an opposite charge bonded to the stationary phase and establish an equilibrium with the mobile phase, for example, the ion exchange chromatography of proteins.

7.7.4 Size Exclusion Chromatography (SEC)

SEC (see Section 7.10) allows for samples with unique size and shape to interact with the porous spaces contained within the stationary phase and establish an equilibrium with the mobile phase, for example, the SEC of proteins.

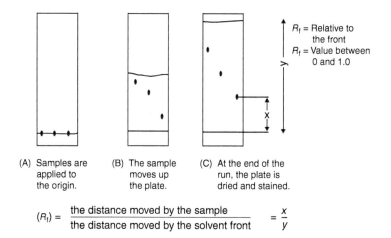

(A) Samples are applied to the origin.

(B) The sample moves up the plate.

(C) At the end of the run, the plate is dried and stained.

R_f = Relative to the front
R_f = Value between 0 and 1.0

$$(R_f) = \frac{\text{the distance moved by the sample}}{\text{the distance moved by the solvent front}} = \frac{x}{y}$$

Figure 7.10 *A typical thin layer chromatography experiment where (a) is the start of the experiment when samples are applied to the stationary phase at the origin before being placed in the mobile phase solvent (b) the samples migrate up the plate, the progression is determined by the samples partition coefficient (Kd) (c) when the solvent front reaches the end of the TLC plate the plate is removed and the developed by a staining process. The distance moved by the solvent front and each sample is noted to determine the Rf value.*

7.7.5 Affinity Chromatography

Affinity chromatography (see Section 7.13) allows for a sample with a unique structural feature to interact with a complementary structure bonded to the stationary phase and establish an equilibrium with the mobile phase, for example, the affinity chromatography of proteins.

7.8 Thin-layer Chromatography (TLC)

TLC is a solid–liquid chromatographic technique. The stationary phase is a micro granular solid that sample components can adsorb to, thus establishing equilibrium with the liquid mobile phase that passes over the stationary phase. The most used solids in this form of chromatography are silica gel (SiO_2), alumina (Al_2O_3) and cellulose. Polar groups on the surface of these solids form weak interactions (H-bonds, van der Waal and dipole interactions) with samples dissolved in an organic liquid. Good solids for adsorption chromatography should be insoluble in the mobile phase, inert to reaction with the sample and have a small particle size to increase the available surface area.

TLC is a sensitive, fast, simple and relatively inexpensive analytical technique. Several samples can be run on the same plate with standards to help in the identification of unknown compounds. It is usually conducted in a planar format with the micro granular solids (adsorbent) bonded to a flat plastic or glass plate. This technique can also be scaled up into a column format to separate larger amounts of sample.

Worked Example 7.3 Thin-layer Chromatography (TLC) of Amino Acids

In a planar format, TLC involves spotting the sample onto the adsorbent near one end of the TLC plate. The plate is placed into a tank containing a small volume of solvent. The tank is covered, and the solvent moves through the stationary adsorbent by capillary action (see Figure 7.10).

- With a pencil lightly score a sample application line (origin) 1 cm from the bottom of the cellulose TLC plate and mark application points on this line.
- Using a capillary applicator spot 10 µl of the amino acid sample onto an origin point. The best results are achieved by taking time to apply a small concentrated spot rather than a large dilute spot. Hold the applicator in one hand and a hot air blower (hair dryer) in the other, then 'dot and dry', 'dot and dry' repeatedly applying a small volume until all the sample has been applied.
- Place the plate in the TLC chamber. The solvent (butanol: acetic acid: water 46:10:26 mixed 50:20 with acetone) level in the chamber should be just below the point of sample application on the TLC plate.
- The solvent will move through the plate by capillary action.
- When the solvent front reaches 1 cm from the top of the plate, remove the plate from the tank and mark the solvent front.
- Dry the plate with a hot air blower and stain the plate with 0.2% (w/v) ninhydrin in acetone (take care to wear gloves).
- Heat the plate with the hot air blower (the ninhydrin reaction with amino acids works best at 70 °C) and record your observations by (a) noting the colour (different amino acids produce different shades of purple and proline produces a yellow colour) and (b) determine the R_f value.

$$R_f = \text{Retention factor} \left(\text{relative to the solvent front} \right)$$

$$R_f = \frac{\text{the distance moved by the sample} \left(\text{centre of the spot} \right)}{\text{The distance moved by the solvent front}} = \frac{X}{Y}$$

7.9 High-Pressure Liquid Chromatography (HPLC)

As mentioned above, the same adsorbent and solvents used in TLC can be used in a column format to separate larger amounts of sample. This is usually done by filling a glass column with a stopper at the bottom with adsorbent, and after applying the sample, the column is eluted with solvent and fractions are collected until the sample elutes.

The particle size of the adsorbent in these columns is relatively large (50 μm), which means there is adequate space between particles for the solvent to percolate through the column under the effects of gravity (hydrostatic pressure). Using these large particles, adequate resolution of complex mixtures is achieved using small volumes of concentrated samples applied to columns with large volumes of adsorbents.

Key to Figure 7.11

A. Solvent reservoirs are usually made from glass. There is usually a filter at the end of the solvent inlet to help prevent particles from entering the chromatography system.

B. High-pressure pumps are required to deliver the mobile phase (at flow rates up to 10.0 ml min^{-1}) through the stationary phase at a pressure up to 6000 psi/400 bar (up to 20 000 psi/1400 bar for ultra-high-pressure liquid chromatography (UHPLC).

C. The sample is normally loaded onto the column using a fixed volume injection loop. Manual or automatic injection systems can be included in a system set-up.

D. HPLC columns come in a variety of different formats. They are traditionally made from steel tubing with an internal diameter (i.d.) of about 4.0 mm, approximately 5–50 cm in length and are normally run at flow rates up to 5 ml min^{-1}. Microbore columns (i.d. 2.0 mm) operate at flow rates up to 0.2 ml min^{-1}, and capillary columns 5.0 cm in length (i.d. 0.2 mm) have nl min^{-1} flow rate. The reduced flow rates from the smaller diameter columns are ideal when interfacing with a mass spectrometer detector.

Silica-based resins are unstable in alkali environments (>pH 8.0). Chemically modified silica resins have been produced to overcome this problem and allow separations to take place at elevated pH values. Microparticulate resins (3–10 μm diameter) are the most popular form of silica stationary phase used in HPLC. These resins can be porous; the smaller the pore size, the greater the overall surface area of the stationary phase. Other chemical modification of silica has resulted in the production of small silica particles

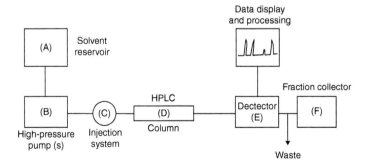

Figure 7.11 *Schematic diagram of an HPLC, Flash or biological chromatography systems.*

(1–2 μm diameter) which are more pH stable and able to withstand elevated pressures. Pellicular resins (30 μm diameter) have a solid core to the particle to which is bonded a thin layer of silica (1–2 mm). This porous outer layer allows rapid interactions to take place between the mobile and stationary phases and is popular in ion exchange resins (see Section 7.9). In the separation of biological molecules, cross-linked dextrans, agarose or acrylamide porous beads are used instead of silica (see Section 7.9.1) and are packed into glass or plastic columns. Mixtures of biocompatible materials with small bead diameters (10 μm) have been manufactured, that are robust enough to withstand high flow rates (>5.0 ml min^{-1}).

E. Online spectrophotometers capable of measuring UV/visible (200–800 nm) light are popular detectors. Other online detectors include fluorimeters, standard, electrochemical and mass spectrometers. The detectors are usually linked to a computer or integrator to facilitate the analysis.

F. Fraction collectors can be included to collect valuable components after they have passed through the detection system. HPLC instruments are primarily analytical systems and may not routinely use a fraction collector, whereas they are essential in the preparative biological or Flash chromatography systems.

To decrease the column size while retaining the theoretical plate number and the resolving power (see Worked Example 7.1), the particle size must be decreased. As the particle size diameter decreases (e.g. 5–2 μm), there is less space for the mobile phase to easily progress from the start to the end of the column. The solvent will no longer percolate through the column due to gravity and requires pressure to force it through the column. In high-pressure liquid chromatography (HPLC), mechanical pumps provide the means to force the mobile phase through the stationary phase (1000–6000 psi: 50–350 bar).

Biological macromolecule chromatography systems, such as AKTA (GE Healthcare UK) Biologic and NGC system (Bio-Rad, UK), are preferentially used instead of an HPLC system to purify biological molecules. The component set-up of such systems is identical to the system set-up in Figure 7.11 with the use biocompatible materials. Polyether ether ketone (PEEK) tubing is used instead of the stainless-steel tubing that would be used in HPLC. The PEEK tubing is flexible, robust and chemically inert to most buffers and solvents. The internal surface is smooth, which helps to prevent sample mixing as the sample flows through the system. Fixed wavelength detectors (260 nm for nucleic acids and 280 nm for proteins) are used rather than the variable wavelength or diode array detectors used in HPLC. Flash chromatography systems (Agilent, Biotage and Interchim Ltd.) use 15–50 μm diameter silica or polymer beads in low- to medium-pressure preparative HPLC systems. Elevated flow rates >5.0 ml min^{-1} allow for the rapid preparation of chemical or biological components prior to analytical analysis using a conventional HPLC. Ultrahigh-pressure liquid chromatography systems (UPLC) use metal columns with an internal diameter of 3.0–10.0 mm and column length of 50–20 mm, which are packed with silica beads with a diameter < 2 μm. The UPLC systems operate using pressures up to 20 000 psi (1400 bar) with 1–25 μl injection volumes and flow rates (<1.0 ml min^{-1}) and can be partnered with a mass spectrometer as a detector.

7.9.1 Normal Phase (NP), Reversed Phase (RP) Hydrophobic Interaction (HIC) and Hydrophilic Interaction Liquid (HILIC) Chromatography

Two common chromatographic separations used with an HPLC system are normal and RP chromatography. Small particles of silica (diameter 5–2 μm) provide the solid matrix base for the stationary phase, to which is bonded a monolayer of a polar material (e.g. alkyl amine) in *normal* phase chromatography or a monolayer of organic material, for example, (C$_{18}$ aliphatic chain) in *reversed* phase chromatography.

In normal phase chromatography, the polar elements of sample molecules will tend to adsorb to polar material surrounding the solid chemistry of the stationary phase when the mobile phase is non-polar, for example, hexane. Different compounds will have different levels of polarity with the most polar being retained on the column longer, increasing the retention time. Normal phase chromatography is most useful when it is used with compounds that are predominately non-polar but contain an element of polarity, for example, phospholipids, steroids and fat soluble vitamins such as vitamin E. Elution can be achieved by increasing the polarity of the solvent, but this can also generate problems, as water molecules can then interact with and deactivate the stationary phase. This results in changes in the retention times of the compounds applied, making the reproducibility of the technique very difficult. The changes in retention time can be accommodated by constantly comparing the retention time of the sample to the retention time of an internal standard (see Section 7.3).

An increasingly popular alternative to normal phase chromatography is hydrophilic interaction chromatography or hydrophilic interaction liquid chromatography (HILIC or HiLiC). This technique was suggested by Alpert (1990) where the stationary phase can be 'bare' non-bonded silica beads using the hydrophilic silanol residues in silica to attract a layer of water molecules. This allows hydrophilic elements of the molecules applied to the column in the mobile phase to partition onto the silica beads. The elution profile in HILIC gradually reduces the percentage of solvent in the mobile phase (e.g. 98–5% (v/v) methanol or acetonitrile) where molecules elute in order of increasing polarity (see Figure 7.4c). Other bonded phases that can be used in HiLiC include amino, amide, cationic and zwitterionic residues. Each of the different HILIC phases allows the preferential partition of a wide range of molecules. The technique of HILIC is particularly useful in the separation of polar molecules which fail to interact with the hydrophobic bonded phases in RP chromatography.

Reversed phase (RP) chromatography is the most widespread form of chromatography because of its versatility. A variety of organic compounds (C_5, C_8, C_{18} and phenyl) can be bonded to the stationary phase to facilitate the separation of many different compounds. In RP chromatography, the initial mobile phase is polar, containing a small percentage (5–10% v/v) of a solvent such as acetonitrile or methanol. When a sample is applied to a RP column, there is a liquid/liquid partition between the polar mobile phase and the non-polar oily 'liquid' monolayer surrounding the silica bead. Elution is achieved by increasing the proportion of solvent in the mobile phase (e.g. 5–80% v/v solvent). The polar compounds elute earlier than the non-polar compounds in the sample. The increase in the proportion of organic solvent in the mobile phase becomes increasingly attractive to any non-polar components in the sample. If an unknown mixture of compounds is to be analysed, the first choice of resin is usually a C_{18} (ODS) phase. This phase has proven to be extremely versatile and reproducible. If an adequate separation cannot be achieved with a C_{18} resin, there are many different manufacturers of RP resins with subtly different chemistries which can be subsequently tested for their ability to facilitate the desired separation.

Hydrophobic interaction chromatography (HIC): Proteins contain amino acids with hydrophobic residues (e.g. leucine, tyrosine and valine) and are normally kept away from water molecules because they contain hydrophobic residues. To minimize their contact with water, these hydrophobic amino acids can be buried in the core of a protein's structure or, if they are near to the surface, they can be covered by hydrophilic or polar amino acid residues. In the presence of high salt concentrations, the surface hydrophobic amino acid residues in a protein's structure can become exposed. These hydrophobic residues in the protein's structure can interact with hydrophobic molecules (e.g. C_8, C_5 or phenyl) bonded to the stationary phase of an HIC resin. The strength of binding depends upon how many and what types of hydrophobic amino acids are involved in the binding; decreasing the concentration of salt in the mobile phase (see Figure 7.4c) will encourage the proteins bound to the column to enter the mobile phase and elute from the column folding again into their normal three-dimensional structure.

HIC works in a similar way to RP chromatography, in that organic ligands bonded to a stationary phase interact with hydrophobic groups on the structure of sample molecules. Proteins will bond to RP stationary phases, but the high ligand density on RP resins means that the interaction is strong, requiring organic solvents or detergents to remove the protein from the RP stationary phase. The proteins will elute, but they will no longer easily fold into their normal biologically active structure; they will be biologically inactive. The lower ligand density on an HIC stationary phase means that proteins can bind to the HIC resin and elute while retaining their biological function.

7.10   Gas/Liquid Chromatography (GLC)

In a similar manner to RP chromatography, GLC involves an organic compound bonded to the stationary phase, effectively forming a monolayer of an organic liquid. The stationary phase is usually packed into a glass capillary column (3 mm i.d.), which is then wound onto a reel and inserted into an oven (see Figure 7.12). The mobile phase is an inert gas (typically nitrogen) pumped through the stationary phase under pressure.

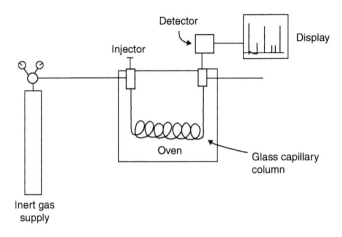

Figure 7.12 *Schematic representation of GLC.*

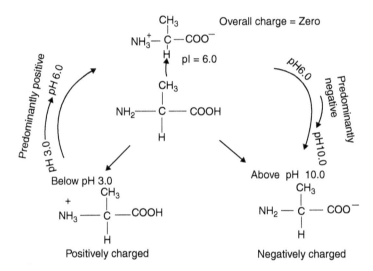

Figure 7.13 *The structure of alanine and the change in overall charge with pH.*

The volatile sample is injected onto the head of the chromatographic column where it can partition into the organic liquid bonded to the stationary phase. The sample is transported through the column by the flow of the inert gaseous mobile phase. The oven housing the stationary phase can be programmed to provide a gradient of heat, gradually increasing the temperature (up to 200 °C) over a period of time. Different compounds volatilize (enter the gas phase) at different temperatures, where they are carried by the mobile gas phase to the end of the column and the detector.

7.11   Ion Exchange Chromatography (IEX)

Ion exchange chromatography (IEX) can be used to separate molecules based on their overall charge because it involves an equilibrium between a stationary charged resin and oppositely charged molecules in the mobile phase (usually involving a buffer to maintain the pH). Ion exchange resins can be divided into positively charged anionic resins which attract anions (negatively charged molecules) and negatively charged cationic

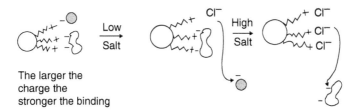

Figure 7.14 *Binding and elution from an anion exchange resin.*

resins which attract cations (positively charged molecules). The basis of the separation in ion exchange chromatography is that different molecules at any pH value will have different levels of charge.

Strong acids readily dissociate in solution at all pH values, whereas the dissociation of a weak acid is dependent on the pH of the solution. Many biological molecules (e.g. amino acids, proteins, nucleic acids and charged metabolites) have weak acid groups in their structures, and the charge on the molecule is dependent on the pH (see Chapter 1, Section 1.6.3.1). For example, amino acids have both positive and negative weak acid groups in their structures (zwitterions). At a low pH, alanine (see Figure 7.13) has an overall positive charge and at a high pH, the overall charge is negative. At pH 6.0, the positive and negative charges on alanine are balanced, making the overall charge on the molecule zero. This is called the isoelectric point (pI); at this point, the molecule will not bind to an oppositely charged particle (ion exchange chromatography) or it will not move under the influence of an electrical field (see Chapter 6, Section 6.3.4). Between pH 3 and 10, alanine has a mixture of positive and negative charge.

Proteins are polymers of amino acids, some of which have weak acid functional groups (aspartic acid and glutamic acid are negatively charged at pH 7.0; lysine and arginine are positively charged at pH 7.0). This means that proteins, like amino acids, have zwitterionic properties. These weak acid charges contribute to a protein's structure, functionality and pI. At 1.0 pH unit below the pI of a protein, it will have sufficient positive charge to bind to a cation exchange resin, and at 1.0 pH unit above its pI, it will bind to an anion exchange resin. The overall level of binding depends on the net charge at that pH. Molecules can be eluted from ion exchange resins by (i) increasing the concentration of salt (e.g. NaCl) in the mobile phase (see Figures 7.4b and 7.14) or (ii) raising the pH to the pI of the molecule of interest (cation exchange chromatography) or lowering the pH to the pI of the molecule of interest (anion exchange chromatography) or (iii) a combination of (i) and (ii).

Silica (porous or pellicular: see the key for Figure 7.11) and polystyrene are used as the solid support in ion exchange resins able to withstand high pressures. Cellulose, dextrans and cross-linked acrylamide are popular supports for low-pressure stationary phase ion exchange resins. The functional groups attached to the resins can be described as weak (e.g. WAX, carboxymethyl or diethylaminoethyl) or strong (e.g. SCX,

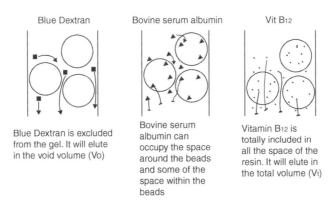

Blue Dextran is excluded from the gel. It will elute in the void volume (Vo)

Bovine serum albumin can occupy the space around the beads and some of the space within the beads

Vitamin B_{12} is totally included in all the space of the resin. It will elute in the total volume (Vt)

Figure 7.15 *The paths taken by three components in size exclusion chromatography (blue dextran ■; bovine serum albumin ▲ and vitamin B_{12} ●).*

SAX, quaternary ammonium or sulfopropyl). This does not reflect the strength of binding but does reflect their useful operating pH range. Strong ion exchange resins operate over a wider pH range than weak ion exchange resins.

7.11.1 The Purification of Antibodies

Antibodies are biomolecules with a high specificity for their target antigen. Their specificity means that they can be utilized in many ways as molecular tools to quantify the amount of an antigen in a cell extract (immunoassays; see Chapter 9) and to observe and locate target antigens in cells using immunofluorescence (see Chapter 2 Section 2.4.4). Antibodies are also used in routine diagnostic test kits, such as those that alert a woman to the production of human chorionic gonadotrophin (hCG) in the early stages of pregnancy. During the global pandemic (2019–2022) caused by variants of the SARS-CoV-2 (covid) virus, lateral flow tests based upon antibody detection were commonplace.

Many of the new approved drugs introduced into the UK National Health Service (NHS) have been humanized monoclonal antibodies. An example of a well-established therapeutic monoclonal antibody is Trastuzumab (sold under the name Herceptin) which is used to treat breast and stomach cancer. The monoclonal antibody Herceptin is raised against the HERP-2 receptor protein, an integral plasma membrane tyrosine kinase, which is over expressed in 20% of breast cancers. Another example is the neutralizing monoclonal antibody (nMab) Sotrovimab, which is used to manage patients with problematic covid virus symptoms. To be used as an academic, diagnostic, or therapeutic tools the monoclonal antibody must be pure and free from any contaminating proteins.

The purification of polyclonal or monoclonal antibodies can be achieved by combinations of several different techniques including salt precipitation, hydrophobic interaction chromatography (see Section 7.7.1), cation exchange chromatography (see Section 7.9) size exclusion chromatography (see Section 7.10) and affinity chromatography (see Section 7.11).

Antibody binding proteins produced by bacteria have been utilized in the purification of different antibody classes (e.g., IgA, IgG, and IgM) from different animal species. Protein A is an antibody binding protein (Mr 49 000) found in the cell wall matrix of *Staphylococcus aureus*, protein G (Mr 65 000) is an antibody binding protein found in groups C and G of *Streptococcal* bacteria and protein L (Mr 95 000) is an antibody binding protein isolated from *Peptostreptococcus magnus*. These bacterial antibody-binding proteins exhibit different binding characteristics with different classes of antibody from different species. For example, protein G does not bind IgA, IgD, or IgM and is the better choice for the purification of mouse IgG_1 and human IgG_3. Protein A shows strong binding to mouse IgG_2 and IgG_3 as well as human IgG_1, IgG_2 and IgG_4 (see Chapter 9, Section 9.1). The use of protein L is restricted to those antibodies with kappa (κ) light chains and its main application is in the purification of monoclonal mouse IgG from ascites fluid or cultured cell supernatant. These bacterial antibody binding proteins have been produced as modified recombinant proteins, which has maximized their antibody binding. Judicial research is required to ensure the correct antibody binding protein is used as an affinity ligand.

The high specificity of antibodies to their target antigen can be utilized to purify the antigen from cell or tissue extracts (immunoaffinity chromatography). An antibody to a target antigen (protein or small molecule) can be covalently linked to a chromatography resin (e.g., activated agarose) with the antigen binding section of the antibody structure pointing into the mobile phase. The partially purified extract can be slowly passed through the column containing the antibody allowing the antibody bound to the column to interact with the antigen in the mobile phase. Following a washing stage to remove any protein that has bound to the column in a nonspecific fashion, the antigen can be eluted (e.g. by dropping the pH to 2.0). The pH is immediately returned to pH 7.0 by including a small volume of concentrated Tris buffer in the fraction collecting tubes. The fractions containing the antigen eluted from the immunoaffinity column are pooled and used for further analysis.

7.12 Size Exclusion Chromatography (SEC)

SEC[3] is a liquid/liquid partition between the mobile phase and the accessible liquid volume contained within the pores of the stationary phase. There is no physical interaction between the sample and the stationary phase, and the separation is also independent of the mobile phase used. Size exclusion can be employed to separate organic and biological polymers (proteins and nucleic acids).

When a sample is applied to a column packed with a size exclusion stationary phase, the molecules will permeate through and around the porous beads in the column. The only parameter that affects the separation is the size and shape of the sample molecules. The pores within the stationary phase have a maximum pore size, which prevents larger molecules from entering the porous network. The molecules in the sample that are too large to enter the pores of the stationary phase are *excluded* from the stationary phase based on their *size* and shape (See Figure 7.15).

Any mixture of molecules applied to a column packed with a size exclusion resin will emerge from the column with the largest molecules eluting first. These large molecules will only be able to move in the mobile phase around the beads (a smaller volume than the total column volume), and they will emerge from the column first. This volume is called the void volume (Vo). Small molecules will also be able to move in the mobile phase around the beads, but because of their small size they will also be able to enter and diffuse into the liquid volume within the beads. Effectively, they move through the column in the total volume of the column (Vt). Depending on their size, other molecules within the sample will be able to access different percentages of the space within the beads. They will elute in an elution volume (Ve) between the volumes of the large molecules (Vo) and the small molecules (Vt). An effective partition coefficient (Kav) can be calculated using these volumes (see Worked Example 7.4).

Worked Example 7.4 Determining the Kav of a Protein in Size Exclusion Chromatography (SEC)

- Blue dextran (Mr 2 000 000), vitamin B_{12} (Mr 1580) and bovine serum albumin were loaded onto a size exclusion column, and their elution volumes were measured (see Figure 7.15).
- Blue dextran eluted in 80 ml, vitamin B_{12} eluted in 240 ml and bovine serum albumin eluted in 160 ml.
- The large molecular mass of the blue dextran means that it will be totally excluded from the pores of the size exclusion gel eluting in the void volume (Vo). The small molecular mass of the pink component (vitamin B_{12}) means that it can access the space around the gel (Vo) and the space within the gel to give the total volume of the column (Vt).

$$Kav = \frac{Ve - Vo}{Vt - Vo}$$

Ve is the elution volume of the sample of interest.
Vo is the void volume of the column.
Vt is the total volume column.

$$Kav(Blue\,Dextran) = \frac{80 - 80}{240 - 80}$$
$$= 0$$

$$Kav(vitamin\,B_{12}) = \frac{240 - 80}{240 - 80}$$
$$= 1.0$$

$$Kav(bovine\,serum\,albumin) = \frac{160 - 80}{240 - 80}$$
$$= 0.5$$

Kav is a number between 0 (Vo) and 1.0 (Vt) without units. It represents a percentage of the stationary gel volume that has been occupied by a component in the sample during the chromatographic run. In this example, bovine serum albumin occupied 50% of the available pore space in the column because of its shape and relative molecular mass (Mr).

- In SEC, the partition coefficient (Kav) of a protein is inversely proportional to the Log_{10} of the proteins relative molecular mass (Mr). Therefore, a column can be calibrated by recording the elution volumes of standard proteins of known relative molecular mass. The Kav can be determined for each protein and a calibration graph constructed plotting Log_{10} Mr on the y-axis against Kav on the x-axis. This calibration graph can then be used to estimate the molecular mass of an unknown protein (also see Chapter 6).
- Size exclusion separates proteins in their native conformation. If a protein is composed of multiple subunits, the total molecular mass of the whole protein assembly will be determined. This is in contrast with denaturing polyacrylamide gel electrophoresis (SDS-PAGE), which estimates the relative molecular mass of a protein's polypeptide chain(s). In denaturing polyacrylamide gel electrophoresis, if a protein is composed of different subunits, these will be separated individually (see Chapter 6).

The stationary phase beads in SEC can be made from different inert porous materials, such as dextrans, acrylamide, silica and synthetic polymers. During their construction, the pore size can be varied, which results in a wide range of resins that can be used to fractionate molecules of different sizes. Many resins have been strengthened by chemical cross-linking, allowing them to be used with high-pressure systems and run at high flow rates.

7.13 Affinity Chromatography

Affinity chromatography is a form of chromatography that is generally applicable to the isolation of components from biological samples. Complex biological polymers such as proteins have their amino acid content derived from their gene sequence. Each protein is synthesized for a biological function, and affinity chromatography exploits the biospecificity of a protein. Affinity chromatography was originally designed to isolate a specific enzyme from a sample mixture using the substrate of the enzyme bonded to a stationary phase (see Figure 7.16).

Enzymes have evolved stereospecific active sites to bind their substrates with high affinity. In theory, only the enzyme in the applied sample will interact with the substrate bonded to the stationary phase. The other components will percolate through the column without binding. The bound enzyme can then be eluted by adding a high concentration of the substrate (high salt or changing the pH is also used) to compete with

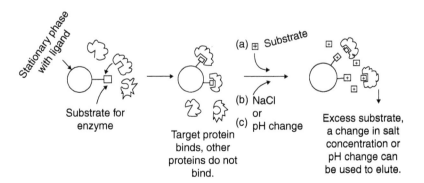

Figure 7.16 *Affinity chromatography of an enzyme.*

and displace the enzyme from the stationary phase. The eluted enzyme can then be collected in a small volume and analyzed. This form of chromatography represents a method with high resolution to quickly isolate biological components from a complex mixture. The interactions between specific nucleic acid sequences and proteins, antibodies and their antigens, cell surface receptors and their agonists have also been successfully exploited in affinity chromatography.

7.14 Summary

- Chromatography is a versatile technique used to separate components of complex mixtures. It is based upon the distribution coefficient (K_D) of compounds between two immiscible phases (i.e. the mobile phase and a stationary phase).
- Chromatography can be used to separate compounds with differences in their polarity, size, volatility and charge using adsorption, size exclusion, affinity, ion exchange or partition chromatography.
- TLC is an inexpensive adsorption chromatography technique for the quick analysis of samples.
- GLC can be used to separate volatile samples
- ODS is a versatile chromatography resin used in RP chromatography. High-pressure pumps are required to deliver the mobile phase through columns containing the small diameter stationary phase particles.
- Proteins can be separated using ion exchange, hydrophobic interaction, size exclusion and affinity chromatography.

Notes

1 The stationary phase may be planar (paper chromatography or thin-layer chromatography; see Section 7.8) or more often contained within a cylindrical steel, glass or plastic column.
2 50% (v/v) methanol is methanol mixed with an equal volume of water, that is, 50 ml of methanol mixed with 50 ml of water (see Chapter 1, Section 1.4.3).
3 The term size exclusion chromatography (SEC) is preferred when the method is used to separate biological molecules (e.g. proteins); this is exactly the same method as gel filtration which is the preferred term when the method is used to separate chemical polymers.

8

CELL CULTURE TECHNIQUES

8.1 Introduction

Bacterial, yeast, plant and animal cells are often grown *in vitro* (i.e. in cell culture medium) to study how they grow, divide or differentiate, also to better understand how they interact with each other and how they respond to various types of mechanical or pharmacological stimuli. Cell culture provides a controlled environment in which cells can be incubated under defined conditions and treated with growth factors, drugs, toxins, or mechanical stimuli in a reproducible fashion. Cultured cells are also widely used in the biotechnology and biopharmaceutical sectors to produce recombinant proteins (e.g. albumin production in bacterial or yeast cell cultures) and antibodies (e.g. in monoclonal hybridoma cell culture technology).

Whatever type of cell is to be cultured, all cell culture methods are based on a common set of principles and requirements:

- The culture should be obtained using a precisely controlled procedure compliant with guidelines for good laboratory practice, or from a recognized cell culture bank, such as the American Cell and Tissue Culture Collection (ATCC: http://www.lgcstandards-atcc.org/).
- Cells need a suitable growth medium containing nutrients to support growth and survival, and this medium must be sterilized before use.
- Cells must be grown under optimized conditions of pH, temperature, humidity, pressure, ionic and osmotic strength.
- Cells must always be handled using an appropriate aseptic technique to prevent the culture from becoming contaminated by unwanted microorganisms.

In the following sections, we will look at basic aspects of aseptic technique applied in culture protocols for the growth of bacterial, animal and plant cells. The main emphasis will be on the handling of bacterial cells, as you are most likely to cover this during the early phase of your studies. As in other areas of laboratory research, accurate dispensing is crucial and general details on pipetting of reagents can be found in Chapter 1. However, this and all other manipulations need to be performed using correct aseptic technique and you are likely to need to use a variety of manual pipetting aids in cell culture work, some of which are shown diagrammatically in Figure 8.1 (see also Chapter 1). Not only will you need to work using aseptic techniques, the materials you use will also have to be sterile. A selection of the most used methods of sterilization and their application are shown in Table 8.1.

ASEPTIC TECHNIQUE IN MICROBIOLOGY

You must *always* follow local health and safety policy for aseptic technique and for safe disposal of waste.

- Correct aseptic technique will prevent accidental contamination of laboratory cultures from external sources (e.g. hands, clothes and environment) and prevent contamination of you and others working in the laboratory.

Basic Bioscience Laboratory Techniques: A Pocket Guide, Second Edition. Philip L.R. Bonner and Alan J. Hargreaves.
© 2022 John Wiley & Sons Ltd. Published 2022 by John Wiley & Sons Ltd.

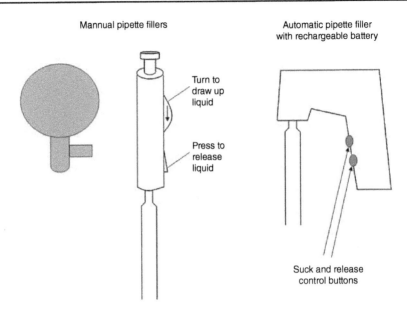

Mannual pipette fillers

Turn to
draw up
liquid

Press to
release
liquid

Automatic pipette filler
with rechargeable battery

Suck and release
control buttons

Figure 8.1 *Diagrams of pipette aids used in cell culture. The reader is also referred to digital images shown in Section 1.5 (Figure 1.5).*

Table 8.1 *Common methods of sterilization of materials and equipment used in cell culture.*

Method	Comments	Application
Heat treatment	Red heat sterilization. This involves heating for a few seconds in a Bunsen flame. This simple and effective method is typically used in microbiology laboratories.	Sterilization of metal implements such as loops, forceps, needles, and so on. Glass rods and spreaders are dipped first in ethanol, which is then burnt off by brief exposure to the flame.
	Baking for 2 h or more at 160 °C or higher (dry heat sterilization). Time consuming.	Sterilization of glassware.
	Autoclaving. This involves heating to about 120 °C, at high pressure for 20 min. It is the method of choice for routine sterilization of most laboratory materials.	Media, glassware, pipette tips and so on. NB: Cannot be used for heat-sensitive media or plastics.
Chemical treatment	The main example is the use of disinfectants.	Used to treat spillages and to disinfect used materials. They need at least 10 min to sterilize contaminated areas.
Filtration	Sterile filters (pore size 0.2 or 0.45 μm) can be assembled in the filter unit and sterilized by autoclaving. They can also be purchased as a one-use pre-sterilized disposable unit.	This is the preferred method of sterilization for heat-sensitive media and reagents.
Radiation	Materials can be sterilized by ultraviolet or ionizing radiation (e.g. gamma rays) but the latter can only be done on an industrial scale.	This method is used by industry to treat most disposable cell culture materials (culture flasks, pipettes, pipette tips, etc.). Materials are discarded after a single use.

- Always wear an appropriate lab coat and wash your hands on entering and leaving the microbiology laboratory.
- Only use sterilized equipment and media, and always decontaminate any used items (e.g. in diluted bleach solution) before discarding them.
- Glass pipettes will be plugged with cotton wool and sterilized in a metal tin by autoclaving. Flame the opening of the tin or wipe the tin with 70% (v/v) ethanol or a disinfectant solution, before removing a pipette to avoid contamination of the remaining pipettes.
- Never place the culture vessel lid (metal, aluminium foil or cotton wool plug) on the bench surface as it will get contaminated or contaminate the bench area. You should remove and hold the vessel lid with the little finger of one hand during manipulations, a technique which becomes easy with practice (see Figure 8.2).
- When working with liquid cultures, pass the open mouth of the glass vessel quickly through a Bunsen flame after removing and before replacing the lid. Working close to the updraught of a Bunsen flame reduces the risk of entry of air contaminants into the vessel (see Figure 8.3).
- Agar-based solid media are prepared in laminar flow cabinets (see Figure 8.4) which enables solid media, which are heated to form molten agar, to solidify by cooling for prolonged periods with the Petri dish lid removed, without risk of bacterial contamination from the air.
- While bacterial cultures can be handled on the bench, a laminar flow cabinet should be used for all manipulations with plant and animal cell cultures.
- Before starting laboratory work, make sure that your bench area is clear of all obstacles, wipe it down with an appropriate disinfection agent (e.g. 70% (v/v) ethanol) and place all items in convenient locations with a clear workspace in front of you.
- Any spillage on the bench should be covered with disinfectant for at least 10 minutes to kill all organisms before mopping up.
- While unlikely to be working with hazardous strains of bacteria at this stage in your career, it is important to be aware of their classification according to the hazard they represent to human health (see Table 8.2).

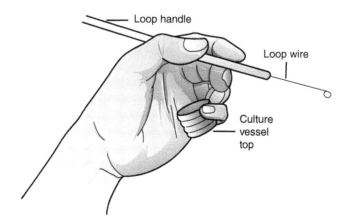

Figure 8.2 *Holding the loop and culture vessel lid during manipulations. The lid of the culture vessel is carefully unscrewed using the little finger, while holding the loop between the thumb and first finger as shown. Always follow local rules on the use of personal protective equipment (e.g. gloves, glasses, etc.).*

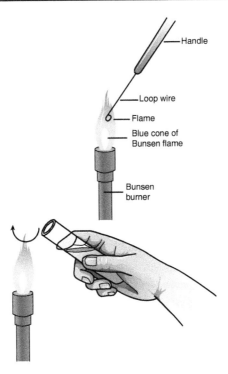

Figure 8.3 *Flaming the loop and the culture vessel opening. The loop wire should be positioned at an angle to the hottest part of the flame, which is just outside the blue cone. When the loop wire is red hot, remove from the flame and allow it to cool for a few seconds before collecting cells. To reduce the risk of contaminants entering the culture vessel, after removing the lid (as shown in Figure 8.2) pass the neck of the vessel briefly through the flame before carrying out each manipulation.*

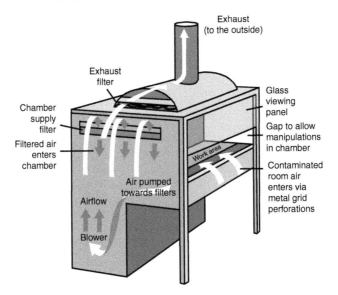

Figure 8.4 *Schematic illustration of the controlled air flow in a laminar flow cabinet. This type of set up is used for preparation of many media for bacterial culture and for most cell culture manipulations with plant and animal cells.*

Table 8.2 *Classification of microorganisms into the appropriate hazard group.*

Hazard group	Likelihood of causing human disease	Groups at risk
1	Unlikely to cause disease.	—
2	May cause disease.	Laboratory workers. Low hazard to the community.
3	May cause severe disease.	May be a serious hazard to laboratory workers and may spread to the community.
4	Causes severe disease.	Serious hazard to laboratory workers and potentially high risk to the community.

8.2 Growth and Maintenance of Cells in Culture

The main focus of this section will be on techniques applied to bacterial culture. However, its applications in eukaryotic cell culture will also be briefly discussed.

As you are no doubt aware, not all bacteria are pathogenic. There are numerous bacteria on exposed surfaces of the body (skin, hair, gut, etc.) that are beneficial to health and prevent invasion by pathogenic bacteria. Indeed, this is a selling point of certain health foods (e.g. probiotic foods and yoghurts). Likewise, there are beneficial and pathogenic strains of bacteria in the environment (e.g. in soil and sewage).

One way to detect beneficial and pathogenic bacteria is to grow samples in bacterial growth media (singular '*medium*'), which may be solid or liquid. Media may be non-selective, selective and/or differential. Non-selective media allow the growth of a range of bacteria, whereas selective media only allow the growth of specific bacteria or groups of bacteria. Differential media can be used to identify a specific subgroup of bacteria, for example, due to a colour difference in growing colonies. Examples of typical growth media and their applications are shown in Table 8.3.

8.2.1 Solidified Media

Many bacteria can be applied to and cultured on the surface of agar-based media. An advantage of this technique is that each single viable cell inoculated onto the agar surface can divide to form a colony of identical cells that is visible to the naked eye. This technique forms the basis of most bacterial isolation and purification methods, including procedures such as 'streak dilution', 'spread plate' and 'pour plate', which are discussed in more detail below. The main component is normally agar, which is melted at high temperature, after which other nutrients are added and the agar is allowed to solidify by cooling it to room temperature.

8.2.1.1 Streak Dilution

This is one of the primary techniques used for obtaining pure microbial cultures. It involves spreading bacteria collected on a metal loop over an agar surface until individual cells fall off the loop in some areas. In theory, each individual cell can divide and give rise to a colony of identical cells. However, as the source material (typically exponentially growing cells in liquid medium) usually contains a very high cell density, a series of steps is required to dilute the applied sample.

Table 8.3 *Selected examples of typical growth media for cell culture.*

Medium	Cell type	Composition[a]
Dulbecco's modified Eagle's medium (DMEM)	Animal cells (mammalian) grown as monolayers or in suspension	Inorganic salts, amino acids, glucose, sodium pyruvate, phenol red pH indicator
Murashige and Skoog medium	Plant cell cultures (various types)	a. Macronutrients (to provide the six major elements: phosphorous, nitrogen, potassium, calcium, magnesium and sulphur) b. Micronutrients (Iron, manganese, zinc, boron, copper and molybdenum) c. Vitamins (e.g. thiamine, nicotinic acid and pyridoxine) d. Amino acids or other nitrogen supplements (e.g. glycine, glutamine, asparagine, arginine, cysteine or tyrosine) e. Growth regulators (e.g. auxins, cytokinins, gibberellins or abscisic acid) f. Sugars (e.g. sucrose or other sugars) g. Undefined organic components (e.g. coconut milk and casein hydrolysate)
Selective enrichment medium	Extreme halophiles (bacteria that grow in extreme salt environments – halobacteria)	High NaCl (about 250 g/l), casamino acids (a source of S, N and P), yeast extract (as a source of growth factors), tri-sodium citrate (as a C and energy source) and various inorganic salts (as a source of K^+, Mg^+, Fe^{2+} and S)
Minimal medium for *Bacillus megaterum*	An example of a basic medium for the culture of a heterotrophic bacterium (i.e. a bacterium that requires an exogenous supply of organic compounds in its growth medium in order to survive)	a. Sucrose (about 1% (w/v); as a Carbon and energy source) b. Various phosphates of potassium and ammonium (as buffer components and a source of K^+, N and P) c. Various inorganic salts (as a source of Mg^{2+}, Mn^{2+}, Fe^{2+} and S)

Shown are examples of growth media for the culture of animal, plant and bacterial cells with a brief overview of their composition.

[a] Components may vary (e.g. serum supplements in many animal cell cultures, agar in solid media formulations, blood in 'blood agar' plates, etc.).

STREAKING DILUTION

As most liquid sources of bacteria will contain between 10^5 and 10^7 bacteria/ml, a loop full of bacteria could contain 10^3 to 10^5 cells, which would form enough colonies to completely cover a 10 cm diameter agar plate. Further steps that are needed to obtain colonies from individual cells are summarized below.

- Sterilize the loop in a Bunsen burner flame for a few seconds (see Figure 8.3).
- While the loop cools, pick up the culture tube or flask (containing exponentially growing cells) with the other hand and remove the cap or plug carefully using the little finger of the hand holding the loop (see Figure 8.2).
- Place the loop into the liquid medium (broth) to collect the cells for transfer.
- Replace the lid on the culture vessel and set it down before lifting the agar plate with the non-dominant hand so that the lid faces upwards (see Figure 8.5).

- Remove the Petri dish lid using the non-dominant hand (see Figure 8.5a).
- Streak the plate (see Figure 8.5b) then put the lid down before re-sterilizing the loop.
- Reorientate the Petri dish at 45–90° to the original position.
- Re-streak from the top end of the first streak so that there is a streak dilution of the original streak applied (see Figure 8.5b).
- Replace the lid again and re-sterilize the loop.
- Lift the lid again and re-streak as before at an angle to the second streak (see Figure 8.5c).

Close the lid and re-sterilize the loop before setting it down.

Several streak dilutions can be performed in this way (see Figure 8.5b–e) before the streaked plate is incubated under appropriate conditions until colonies appear (see Figure 8.5f).

NB: Always remember to label Petri dishes clearly with your name, the name of the bacterial strain, and so on, which should be written on the bottom plate so as not to obscure visual examination of the colonies.

8.2.1.2 Spread Plate Dilution

In this method, a small volume (50–500 μl) of serial dilutions (see Chapter 1) of cells in liquid culture medium or an appropriate sterile diluent is applied to the surface of solid agar medium in a Petri dish. It is then evenly spread over the whole agar surface using an L-shaped glass spreader. Prior to this, the spreader is sterilized by dipping it into a beaker containing 70% (v/v) ethanol, the excess ethanol is allowed to drain and the remainder evaporated off by gently passing the rod through a Bunsen flame. Once cooled, it can then be used to spread the cell suspension, as indicated in Figure 8.6. The Petri dish lid should be removed for the minimum time possible and breathing onto the plate should obviously be avoided.

This is one of the main methods used to quantify the number of viable cells or colony-forming units (CFUs) in a bacterial culture sample. It is typically done using a range of dilutions until an easily measurable colony density of between approximately 20 and 200 per plate is obtained. If this requires dilution, multiplication by the dilution factor will give the actual viable cell count as CFUs per mL of original culture (see Worked Example 8.1).

Worked Example 8.1 Using the Spread Plate Technique to Determine Viable Cell Counts

A volume of 0.1 ml of serial 10-fold dilutions of a bacterial culture were applied to Petri dish culture plates and incubated at 37 °C for 24 hours. The neat and 1/10 dilutions had too many colonies to count. The 1/100 dilution produced 165 colonies. The 1/1000 dilution produced only 15 colonies and no colonies appeared at further dilutions. Estimate the number of viable cells per ml of the original culture medium.

- Assuming that each colony arose from a single viable cell, take the colony count for the 1/100 dilution and multiply by the dilution factor to obtain the number of viable cells in 0.1 ml of culture.

$$\text{CFUs in } 0.1 \text{ ml} = 165 \times 100 = 16\,500$$

Multiply by 10 to obtain the viable cell number per ml

$$\text{CFU/ml} = 16\,500 \times 10 = 165\,000$$

Applications:
- This type of approach can be used to estimate the viable cell counts for different types of cultures obtained from clinical isolates, hospital equipment, environmental samples (soil, air and water), and so on.
- If performed using selective media, it can aid in the identification process.
- It can also be used to study the ability of antibiotic molecules to inhibit colony formation.

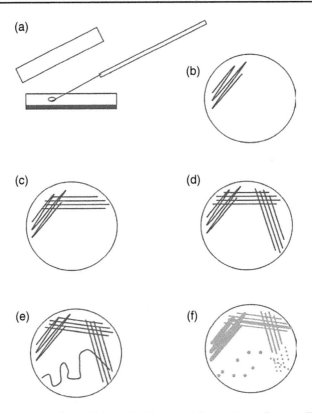

Figure 8.5 *Streaking an agar plate with bacteria. Exponentially growing cells are collected in a loop and streaked onto an agar plate as directed in the Practical hints and tips – streaking dilution information box (a–e). After a suitable incubation period, many colonies appear, some of which are individual (f). Always follow local rules on the use of personal protective equipment (e.g. gloves, glasses, etc.)*

8.2.1.3   Pour Plate Dilution

In this procedure, a known volume of cells grown in liquid medium are mixed with molten agar medium, usually in screw-cap bottles containing a volume sufficient to pour one agar plate (15–20 ml). The agar will only remain molten at temperatures above about 45 °C and, although most bacteria are not killed as a result of a few minutes exposure at 45–50 °C, the procedure can be damaging to bacteria taken from low-temperature environments. The agar is maintained in its molten form in a water bath. The cap is removed, and culture medium added using the aseptic techniques discussed earlier. The cap is immediately replaced, and the cells carefully mixed with the molten medium, which is then quickly poured into the Petri dish and allowed to solidify by cooling with the lid on. The plates are then incubated under appropriate conditions to allow colony formation and counting.

POUR PLATE DILUTION

- Cells should be mixed with the molten agar medium by rotating the bottle between the palms of the hands, as shaking can cause frothing of the medium.
- Colonies which are much smaller than those observed on a streaking dilution plate will form within the solid medium rather than on its surface, giving the advantage that a much larger number of colonies can be counted than with streaking dilution.

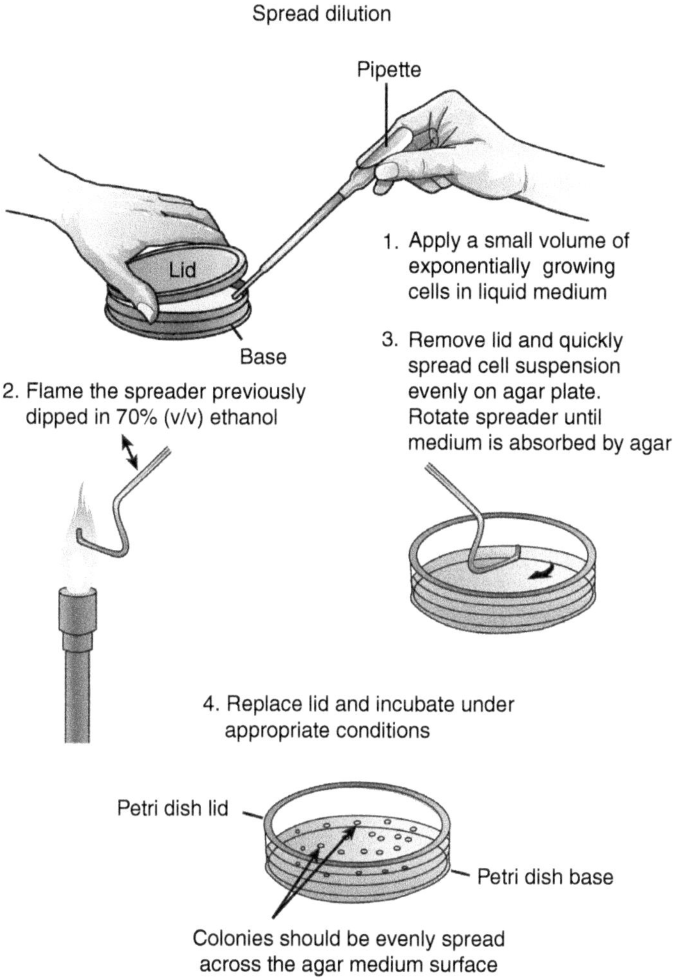

Spread dilution

Pipette

1. Apply a small volume of exponentially growing cells in liquid medium

Lid

Base

3. Remove lid and quickly spread cell suspension evenly on agar plate. Rotate spreader until medium is absorbed by agar

2. Flame the spreader previously dipped in 70% (v/v) ethanol

4. Replace lid and incubate under appropriate conditions

Petri dish lid

Petri dish base

Colonies should be evenly spread across the agar medium surface

Figure 8.6 Spread plate dilution of bacterial cultures. Exponentially growing cells are applied by pipette onto the surface of the solid agar medium and spread using a glass spreader as indicated in the figure itself. Always follow local rules on the use of personal protective equipment (e.g. gloves, glasses, etc.).

- As these colonies are also different in shape to those observed on a streaking dilution plate, it is often the case that an additional thin layer of molten agarose is placed on top of the previously poured medium to prevent the growth of wider colonies on the medium surface.
- Remember to disinfect or sterilize the glass rod before setting it down.

8.2.2   Other Applications of Solid Media Culture Methods

8.2.2.1   Bacteriophage Detection

Bacteriophages ('phages') are viruses that infect bacteria. Bacteriophage-mediated lysis of bacteria grown on agar plates is often used in laboratory classes to demonstrate the detection of viruses and the determination of their quantity in an extract or isolate. Although phages are too small to be seen using a light microscope, their presence in a bacterial culture can be detected by measuring their ability to cause lysis (destruction) of bacteria grown on agar plates.

For this reason, bacterial cells cultured in liquid medium are dispersed in molten agar with a known number of bacterial phages, as described above. This is then poured out to form a thin layer on the surface of a layer of the same agar in a Petri dish. If no phages are present, a continuous 'lawn' of bacteria grows throughout the top agar layer following incubation under appropriate conditions.

When virulent phages are present, clear circular areas termed 'plaques' are observed in areas where a phage is present (see Figure 8.7). Each plaque is the result of a single infective phage. Virulent phages are capable of infecting a cell, replicating within the cell and causing cell lysis within an hour. On lysis, more phages are released to infect neighbouring cells growing in the surrounding agar medium. As the lytic cycle of virulent phages is often very short, plaques need to be counted within a short time window as they get larger over time and may begin to fuse with each other making accurate counts difficult to achieve.

Counting can be facilitated by viewing the plate against a dark background and marking the plaques on the base of the plate with a marker pen. The number of plaque-forming units (PFUs) per ml of virus culture can be determined in a manner similar to that described earlier for bacterial CFUs, taking into account the volume and dilution (if any) of viral extract at the point of application to the molten agarose medium (see Figure 8.7).

8.2.2.2 Plant Tissue Culture

Plant tissue can also be cultured on agar by removing small areas of tissue from a plant (explants), sterilizing their surface (e.g. by rinsing for a few minutes in a solution of 10% (w/v) sodium hypochlorite), after which it is transferred to an agar-plate containing a suitable growth medium to support the growth of a mass of undifferentiated cells or 'callus' (see Figure 8.8). Using an appropriate combination of growth medium and plant hormones, it is possible to stimulate the growth of new plantlets. Such approaches can be used to determine the requirements for the regulation of plant growth and development in experimental work. They can also be used to produce infection-free stocks of rare, endangered or commercially important plants to

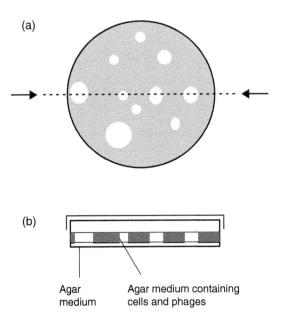

Figure 8.7 *Bacteriophage detection on agar plates. Shown is a schematic diagram of an aerial (a) and cross-sectional view (b) showing the appearance of clear plaques where active phages are present in a bacterial culture on solid medium.*

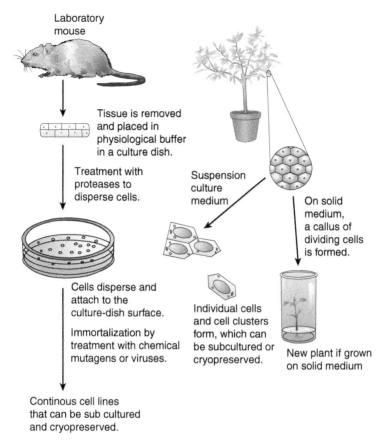

Figure 8.8 *Typical workflow for the generation of plant and animal cell cultures. Note that certain animal cell types (e.g. white blood cells) are grown in suspension culture.*

allow their reintroduction to the environment at a later date. Plant cells derived from callus tissue can also be transferred to liquid culture (see Figure 8.8).

8.2.3   Liquid Media

In situations where the rate of growth is to be determined (e.g. by constructing a growth curve) or there is a requirement to generate large cell numbers (e.g. in the large-scale production of recombinant proteins), cells can be grown in a suitable liquid medium usually with agitation. Again, if you come across this technique in classes, you are most likely to be handling bacterial cell cultures and hence the main emphasis in this section will be on microbiological examples, but with some discussion of liquid culture of eukaryotic cells.

8.2.3.1   Cell Suspension Cultures

In microbiological applications, bacterial cells are typically grown in sterile conical flasks in a variety of sizes (200 ml to 2 litres) depending on the cell numbers required. Flasks are rotated on an orbital shaker operating at the required speed (usually 20–250 rpm) and temperature, which depend on the properties of the culture being grown. Larger vessels are often referred to as fermenters or bioreactors in the field of biotechnology.

If the organism being cultured is an aerobic bacterium, it is important to limit the medium to a maximum of 1/5th of the flask volume in order to provide adequate surface area for absorption of air. Larger scale cultures may require more vigorous stirring (e.g. using a sterilized magnetic stirrer) and the introduction of sterile air through a filter system. The simplest air filter is a sterile glass wool or cotton wool plug placed in the neck of the flask. Alternatively, a commercially available filter unit can be used, which would normally have a pore size of 0.2 μm diameter. However, it is also important to note that cultures of anaerobic bacteria are normally grown without agitation to reduce aeration of the cells.

Batch culture is the most common liquid cell culture method and is widely used in small-scale laboratory work. Using aseptic technique, cells are inoculated into a vessel containing appropriate growth medium after which they increase in number following a predictable pattern with four main phases, referred to as lag (slow growth), log (exponential growth), stationary (no net growth) and decline (death), as indicated in Figure 8.9.

Cell growth can be measured in a number of ways including counts in a haemocytometer chamber (see Chapter 2), in an electronic particle counter (e.g. a Coulter® counter), by taking turbidity measurements of the cell suspension in a spectrophotometer (i.e. optical density or attenuance of light; see Chapter 3), and so on, as outlined in Table 8.4. For any cell culture, it is possible to estimate the growth rate constant or generation time according to the equations below (see Worked Example 8.2):

$$\text{Growth rate}(\mu) = \frac{2.303(\log N_2 - \log N_1)}{t_2 - t_1}$$

N_1 and N_2 represent cell counts determined at an early (t_1) and late (t_2) time point on the exponential part of the curve. If times are measured in hours, the units of μ would be expressed 'per hour'.

The time it takes for a cycle of cell replication referred to as the doubling time (T2) or generation time (g) is given by the equation below:

$$T2(g) = \frac{0.301(t_2 - t_1)}{\log N_2 - \log N_1}$$

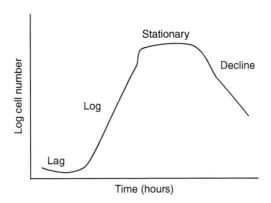

Figure 8.9 *Typical growth curve for cultured cells. All cell cultures grow at their own characteristic rates but usually they conform to a specific pattern of growth involving a growth curve with a lag period followed by rapid growth, stationary and decline phases. Source: Adapted from Alberts et al. (2017). Molecular Biology of the Cell 5th Edition. Garland Press.*

Table 8.4 *Methods of measuring cell growth.*

Methods	Principles	Disadvantages
Microscopy counts	A cell count is made on a diluted cell suspension in a haemocytometer chamber (see Chapter 2).	Time consuming if multiple counts are required. Living and dead cells may be difficult to distinguish. Clumping of cells may affect accuracy. Viable cell counts may be possible using non cell permeable DNA binding dyes (e.g. SYTOX green and propidium iodide) to detect damaged cells. However, an instrument with fluorescence detector is required.
Electronic particle counting	Coulter counters detect changes in electrical resistance as cells pass through a channel, allowing very rapid measurements of total cell counts.	It may be difficult to distinguish living, and dead cells, cell clumps and other particles. A sophisticated alternative is to use a flow cytometer which can sort particles as well as count them.
Cell culture-based methods	Cell cultures work on the assumption that under appropriate conditions, an individual viable (living) cell can multiply to form a colony on agar plates or a measurable change in turbidity (cloudiness) in liquid culture. The main advantage of this approach is that only viable cells are counted.	The true viable count will be underestimated if not all cells are able to grow under the conditions used. Cell clumping and dilution errors can lead to further inaccuracies. Sterile media and equipment are needed. Long incubation times of up to 3 days may be required to produce measurable growth.

Worked Example 8.2 Growth Rate and Doubling Time

If you counted 4000 cells at the first time point and 128 000 cells after 2 hours, the growth rate is calculated from the equation above as:

$$\mu = \frac{2.303 \times (\log 128\,000 - \log 4000)}{2}$$

$$= \frac{2.303 \times (5.11 - 3.60)}{2}$$

$$= \frac{2.303 \times 1.51}{2}$$

$$= \textbf{1.74 per hour}$$

This indicates the number of times per hour that the population will double during exponential growth. The *doubling time* would be calculated as follows:

$$g = (0.301 \times 2) \div (\log 128\,000 - \log 4000)$$

$$= \frac{0.602}{1.51}$$

$$= \textbf{0.4 hours} \,(24 \text{ minutes})$$

Therefore, the population doubles every 24 minutes during the exponential growth phase.

8.2.3.2   Alternatives to Batch Culture

Continuous culture can be used when cells need to be maintained in the exponential growth phase for long periods of time. It involves continuously removing a fixed volume of culture and replacing it with new culture medium so that cells are diluted, and the remaining cells have sufficient nutrients to maintain exponential growth. Any increase in cell number due to proliferation must be offset by a similar loss of cells due to the dilution with fresh medium. While it is useful to be aware of this approach, it is unlikely that you will be exposed to this technique in the early stages of your studies.

Continuous cultures are more difficult to maintain and are more prone to contamination than batch cultures, as they need rigorous aseptic technique and the main culture vessel is connected to at least two additional vessels, one to introduce the culture medium and the other to collect the culture that is removed. However, it has the advantage that a large number of cells at the same stage of growth can be maintained over long time periods, making them useful for physiological and biochemical studies. The process can be automated (computer controlled) and is often used to determine conditions for continuous culture in larger scale bioreactors used in the biotechnology industry (e.g. for the production of antibodies or recombinant proteins).

8.2.4   Liquid Culture of Eukaryotic Cells

8.2.4.1   Suspension Culture

Cells from a range of plants and animals have been successfully grown in suspension culture. However, extensive cell divisions in culture can lead to genetic drift. Procedures are similar to those used for handling bacterial cells except that aseptic manipulations for small-scale cultures are normally carried out using a laminar flow cabinet (see Figure 8.4), which provides an aseptic environment using a filtered air circulation system designed to minimize the risk of microbial contamination.

A wide range of plant cell cultures has been established from different plants and plant tissues. Typically, a fragment of callus tissue (obtained as described in Section 8.2.3) is transferred under aseptic conditions to liquid growth medium in a conical flask, which is then placed in an orbital shaker set at a speed that encourages cell dispersal to form a population containing a mixture of individual cells and cell clusters. A homogeneous culture of individual cells is unlikely to be obtained due to the formation of plasmodesmata between many plant cells.

Certain animal cells such as white blood cells are also grown in suspension culture though the majority of animal cell cultures involve cells grown in the form of a monolayer attached to a plastic culture dish surface (e.g. fibroblasts, epithelial and endothelial cells, neurons, etc.).

8.2.4.2   Monolayer Culture of Mammalian Cells

Most animal cell types are cultured using sterile plasticware, in which the internal plastic surface of the culture dish or 'flask' is usually treated chemically or by microwave irradiation to produce a surface that increases cell attachment to it. If the cells are mitotically active and are left to incubate for too long, the monolayer becomes confluent (i.e. the cells cover the whole culture dish surface) and cells can become detached and/or lose viability as nutrients in the medium become limiting. To avoid this, they are usually 'fed' every couple of days, which involves replacing some of the spent medium with fresh medium. Before becoming confluent, the monolayer is detached from the culture dish surface by brief incubation with a peptidase such as trypsin (see Chapter 5, Section 5.2.1), which degrades proteins that enhance cell attachment to the culture dish surface. The cells are gently removed by pipette to form a suspension, a fraction of which (usually between a tenth and a third, depending on how rapidly the cells divide) is diluted in fresh growth medium. This process of subculture is often referred to as a '*passage*'.

In some studies, the cell culture dish needs to be coated with a protein found in the extracellular matrix (e.g. collagen, fibronectin or laminin) in order to simulate *in vivo* environments more closely. Growing cells in an environment where there is no cell culture-treated surface or extracellular matrix protein coating can encourage the formation of cellular aggregates or spheroids by cells that would otherwise grow as monolayers attached to the culture vessel surface. These three-dimensional aggregates are thought by many to resemble the cellular interactions in tissues more closely than 2D monolayer culture.

As an additional measure against contamination, antibiotics such as penicillin and streptomycin are often added as supplements to culture media. However, this does not protect from contamination by fungi (e.g. yeasts). Nor does it prevent contamination by mycoplasma, a specialized type of bacteria that lack cell walls and are therefore resistant to β-lactam antibiotics (e.g. penicillin). Indeed, there is no substitute for good aseptic technique to minimize the risk of microbial contamination together with regular checks for contamination by microscopic or molecular analysis.

Typical signs of microbial contamination include increased turbidity of the culture medium due to over-growth of yeast or antibiotic-resistant bacterial cells and/or (if phenolphthalein is present in the medium) a change in colour of the medium from red to yellow, indicating reduced pH. Mycoplasma contamination is more difficult to detect by visual inspection. These bacteria are thought to infect around 30% of cell cultures and the first sign of contamination may be subtle changes in cell growth or differentiation patterns and/or altered behaviour in experiments. Regular mycoplasma testing is recommended, which can be achieved, for example, using commercially available PCR kits. If contamination is detected, although mycoplasmas specific antibiotics could be used to remove the contamination, this could cause proteomic or genomic changes in the mammalian cell line. Therefore, if there are sufficient cryopreserved stocks of uncontaminated cells, it is recommended that mycoplasma-contaminated cells are destroyed, and new stocks thawed to continue with experimentation.

As for bacterial cell culture, all equipment and materials must be sterilized prior to their use in eukaryotic cell culture.

Various types of mammalian cell cultures are used in research and in the biotechnology industry. Important features about the main types of mammalian cultures and their production are summarized in Table 8.5 and Figure 8.7.

8.2.5 Seeding of Cells for Experiments

Only healthy exponentially growing cells should be used in experimental work, as overgrowth of cells causes stress, which may change cellular responses in subsequent experimental work. If cells are grown as monolayers, they are typically detached using the peptidase trypsin, then collected by centrifugation and the pellet resuspended in an appropriate volume of fresh culture medium. By contrast, suspension cultures may be used directly. Cells are applied to a haemocytometer chamber using Trypan Blue reagent to monitor cell viability (see Chapter 2, Section 2.4.6.4). Cells should only be used in experiments if they exhibit accept-ably high levels of viability (typically >90%). After seeding, a consistent recovery period should be allowed to allow the cells to adapt to their new conditions prior to experimental manipulations.

Experimental conditions should be pre-optimized to avoid overgrowth of cells during the incubation times required. A consistent cell density per unit surface area or volume should be used for monolayer and suspension cultures, respectively. Cells are then incubated under the appropriate conditions and their behaviour or treatment responses monitored using appropriate endpoint measurements. For example, there are many assays of cell viability, cell proliferation, differentiation, metabolic activity or cell function that can be carried out on cell monolayers or suspension cultures. Alternatively, it may be of interest to specific molecular markers or changes in cell signalling pathways or the levels of second messengers such as cAMP or Ca^{2+} that regulate cell growth and survival.

In some cases, these assays can be monitored *in situ* in living cells. Examples include the use of fluores-cent probes that bind to intracellular Ca^{2+}, thus facilitating the monitoring of changes in Ca^{2+} homeostasis, and the use of membrane permeable substrates to monitor enzyme activity *in situ*. In other cases, the cells may need to be fixed or lysed for further analysis. For example, fixed cell monolayers can be studied for morphological changes and/or used for immunofluorescence staining with antibodies that recognize molecular targets of interest (see examples in Chapters 2 and 9). Cell lysates could be used to measure cell protein concentration (e.g. using the BCA assay – Chapter 4, Section 4.4.4) or to measure the activity of an enzyme of interest after specific incubation conditions have been applied.

Table 8.5 *Selected examples of animal cell cultures.*

Cell line	Source	Applications
Hybridoma	Fusion of a lymphocyte with a myeloma (a type of tumour) cell.	Production of monoclonal antibodies.
Fibroblasts	Connective tissue or skin.	Studies of fibroblast function in tissue organization and wound healing. These are normally finite cell lines (i.e. they survive 40–50 divisions) but transformed (immortalized) cell lines are also available.
Keratinocytes	Epidermis (The major cell type in the outer layer of the skin).	Useful for studies of skin cell function, wound healing and for the creation of artificial skin. These may exist as finite or immortalized cell lines.
Lymphocytes	White cells from the blood.	These can be isolated freshly from blood but used immediately to study T cell functions. Transformed (tumour) cell lines are also available.
Hepatocytes	Liver (The major cell type).	Useful for studies of liver cell function. Primary cultures are short-lived but established (immortalized) cell lines also exist.
Neuroblastoma	From neuronal tumours in the central nervous system.	Primary cultures of neurons are short-lived. Established (tumour) cell lines such as neuroblastoma divide indefinitely and can be used in studies of neuronal cell function. They can be induced to differentiate into a mature neuronal phenotype by appropriate manipulation of the growth medium – useful in studies of development.
Glioma	From glial tumours in the central nervous system.	Primary cultures of glial cells can be obtained from neural tissue, but these are normally short-lived. Glioma cell lines are immortal and can be used in studies of glial cell function and differentiation (development).

Shown is a selection of the types of cell that can be grown in culture. The list is not exhaustive and does not include more recent developments in stem cell cultures. Many cell types are available commercially. These are kept in cryopreservation vessels and dispatched to investigators on dry ice.

8.2.6 Cryopreservation of Cells

A major problem with cultured cells is that, if they are subcultured repeatedly over long periods of time, genetic drift can occur. This may lead to changes in cell phenotype and physiology. It is therefore necessary to cryopreserve stocks of cells after low numbers of subculture, in order to preserve the original features of the cell line. The most effective way of doing this is to gradually freeze exponentially growing cells in growth medium containing cryopreservation agents such as dimethyl sulfoxide (DMSO) at 5–10 % (v/v). Cells frozen in this way can be stored indefinitely at −196 °C to −150 °C in liquid nitrogen or in its vapour phase. When needed, cryopreserved cells are rapidly thawed and, under aseptic conditions, returned to normal growth conditions.

8.3 Summary

- Cells can be isolated and cultured from a range of sources in liquid or on solid media.
- In culture, cells can be used to study a range of physiological phenomena and the effects of toxins or therapeutic agents.
- They can be used on a larger scale in the production of recombinant proteins and/or drugs.
- It is essential that good aseptic technique is used when handling cultured cells and that stocks are cryopreserved at an early stage so as to avoid genetic drift.

Notes

As bacterial cultures grow very rapidly, they are less likely to become contaminated by other bacteria than are animal and plant cell cultures, which are much slower growing.

9

ANTIBODY-BASED ASSAYS (IMMUNOASSAYS)

9.1   Antibody Structure and Uses

Immunoassays utilize antibodies to detect a wide range of target molecules (antigens) against which they have been raised. Antigens range from macromolecules (e.g. proteins, lipids and DNA) to small molecules (e.g. glucose and cAMP). Thus, immunoassays can be useful in a range of applications in basic research, diagnostics and therapy. The suitability of antibodies to be used in this way is down to their structure. The basic unit of antibody structure consists of four polypeptides chains, comprising 2 heavy chains (H) and 2 light chains (L), forming a Y-shaped structure, as shown in Figure 9.1. Most of the protein sequence in antibody molecules is conserved forming so-called constant domains. The antigen binding sites (paratopes) are located within variable domains comprising 110–130 amino acids at the upper tips of the Y-shaped molecule. These sequences are unique to each antibody and are responsible for its specificity at the antigen binding site. Antibody–antigen affinity is typically determined by the dissociation equilibrium constant (K_D), lower values being indicative of higher affinity. The K_D range for typical antibody/antigen binding is in the low nanomolar (1×10^{-9}) to picomolar (1×10^{-12}) range.

An antibody molecule is made up of heavy and light chains, both of which are composed of several homologous subdomains of about 110 amino acids. Each domain contains intra-chain disulphide bridges forming a structural loop containing about 60 amino acids. Light (L) chains contain one variable (V) and one constant (C) domain, whereas heavy (H) chains contain 1 V and 3 or 4 C domains depending on the antibody class. Within the antibody variable regions are areas of much higher levels of variability, which are called hypervariable domains. The latter are interconnected by less variable sequences known as framework regions (FRs).

The biological activity of antibody molecules is determined by their constant domains. As the constant domain is crystallizable, it typically called the Fc domain and forms the tail of the Y which are produced in different types or at different stages of immune responses; these are called IgM, IgG, IgA, IgE and IgD, and their attributes of which are summarized in Table 9.1.

The heavy variable (HV) regions also contain the antigen binding sites in specific areas called complementarity determining regions (CDRs), as their shape is complementary to the epitope (antigenic determinant) recognized. Framework regions act to form a β-sheet structural scaffold that holds the HV regions in position, from which the HV regions project to bind to the antigen. Differences in the amino acid sequence of HV regions create distinct shapes at the tips of the Y-shaped antibody structure, leading to different specificities for antigens.

Studies using X-ray crystallography have shown that each domain in the antibody structure is folded in a characteristic manner producing a structural feature known as the immunoglobulin fold. However, it should be noted that antibodies are not rigid structures, being able to move their Fab 'arms' and Fc 'tail' regions. Further structural change is brought about by the presence of glycosylation sites in the C domain (Figure 9.1).

These unique characteristics of antibody molecules make them useful in a wide range of scientific and medical applications, as summarized in Table 9.2. Antibodies are widely used as tools for research and therapy due to their high level of specificity and flexibility. If the antigen targeted is a protein, antibodies can recognize a variety of features including parts of primary sequence of proteins, chemical changes that regulate protein function (e.g. phosphorylation, acetylation and oxidation) and conformation-dependent epitopes. Antibodies are very useful tools in bioscience and biomedical research due to their high level of specificity and their amenability to be used in a range of ways, as summarized in Table 9.2. They are also used as tools for diagnosis and treatments of diseases; recent examples include the use of antibodies for

Basic Bioscience Laboratory Techniques: A Pocket Guide, Second Edition. Philip L.R. Bonner and Alan J. Hargreaves.
© 2022 John Wiley & Sons Ltd. Published 2022 by John Wiley & Sons Ltd.

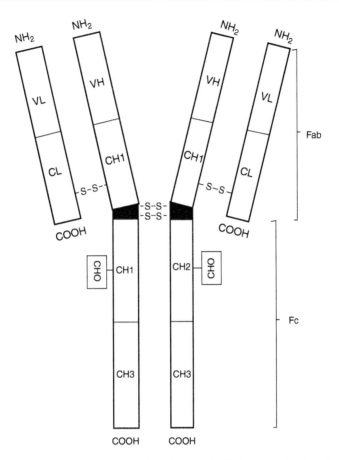

Figure 9.1 *Generalized structure of an antibody molecule. The Y-shaped antibody molecule comprises 2 heavy chains and 2 light chains. Two intermolecular disulphide bridges (-S-S-) link the 2 heavy chains together, while a single disulphide bridge links each light chain to one of the heavy chains. Each heavy chain is glycosylated (CHO). The amino and carboxy termini of each polypeptide chain are indicated by NH₂ and COOH, respectively. Each heavy chain and light chain are made up of constant domains (CH, CL) which show high levels of sequence homology, and variable domains (VH, VL) which show significant variability in their primary amino acid sequences. The Fab fragment contains the antibody binding domains close to the amino terminal ends of the variable regions. The Fc fragment is responsible for the effector functions of each antibody class. The shaded area represents the flexible hinge region in the heavy chain sequence.*

(i) the detection of pathological hallmarks of neurodegenerative diseases, autoimmune disease and cancer, and (ii) for the treatment of cancer and Covid-19.

9.2   Antibody Purification, Labelling and Detection

Depending on the application, antibodies may require purification. Key sources of antibody are serum prepared from animals immunized with the antigen of interest, monoclonal hybridoma culture supernatant containing monoclonal antibodies and ascites fluid from the peritoneal cavity of animals injected with hybridoma cells producing monoclonal antibodies. For many research purposes, the first two of these may be used directly if sufficient cross-reactivity with antigen can be achieved. However, there are impurities in all of the above and enrichment may also be required if the antigen-specific antibody concentration is very

Table 9.1 *Characteristics of immunoglobulins. The main types of immunoglobulins are shown, together with general details about their principal functions and chain structure.*

Type	Properties	Structure
IgA	Found in bodily secretions such as mucous, saliva, tears and maternal milk. Protects against pathogens.	Dimer of 2 immunoglobulin subunits linked end-to-end via their Fc domains.
IgD	This Ig forms the B cell receptor and is thought to modulate the maturation and activation of B cells.	Single immunoglobulin unit.
IgE	This class of Ig plays an important role in mediating the allergic response.	Single immunoglobulin unit.
IgG	This is the most common class of Ig, being secreted in large amounts by antibody-producing cells into the blood. It can cross the placental barrier and be ingested via maternal milk, thus transferring immunity from mother to foetus or child.	Single immunoglobulin unit.
IgM	This class of Ig is produced at an early stage of the immune response.	Pentameric structure comprising 5 immunoglobulin units.

Table 9.2 *General uses of immunoassays. Shown are some of the main uses of immunoassays in research, diagnosis and therapy. Common detection systems are given for each application, though these are not exclusive.*

Application	Format	Common detection systems	Potential clinical applications
Immunoassay-based antigen detection in bioscience and biomedical research	Enzyme-conjugated primary or secondary antibodies can be used to detect cross-reactivity of primary antibody with (i) antigen in enzyme-linked immunosorbent assays of cell and tissue extracts. (ii) On immunoblots of cell and tissue extracts.	(i) Detection of a coloured reaction product on incubation with enzyme-specific substrates. (ii) Release of chemiluminescence in a reaction coupled to enzyme activity (enhanced chemiluminescence – ECL) on Western blots.	Detection and quantification of biomarkers of disease in blood or tissue samples. May be of diagnostic or prognostic value.
Immunocytochemical staining	HRP- or fluorophore-conjugated primary or secondary antibodies can be used to detect cross-reactivity of primary antibody with antigen on fixed cells and tissues.	Chromogenic reaction detected by brightfield microscopy (HRP) or fluorophores by fluorescence microscopy.	Detection of biomarkers of disease in cultured cells, blood samples, tissue biopsies or in vivo. May be of diagnostic or prognostic value.
Therapy	Purified antibodies or antibody fragments directed at a molecular initiating event or a key event in a specific disease. In some cases, an immunization strategy may be used.	Not applicable.	Treatment of a range of diseases including cancer, diabetes and Covid-19.

161

Table 9.3 *Antigens targeted by antibodies. Shown are selected examples of the ways in which specific antibodies can be used to detect protein levels or chemical modifications to proteins of importance in important cell functions. The list is not exhaustive.*

Antibody targets (selected examples)	Examples	Antigen distribution
Proteins	Anti-tubulin, anti-actin, anti-intermediate filament proteins.	Microtubules, microfilaments, intermediate filaments networks in cells.
	Anti-succinate dehydrogenase.	Mitochondrial inner membrane.
	Anti-histone antibodies	Chromatin network in the nucleus.
Acetylated amino acids	Anti-acetylated lysine.	Detects proteins that contain acetylated lysine (e.g. histones in nuclear chromatin and tubulin in stable microtubules in the cytoplasm).
Phosphorylated amino acids	Anti-phosphoserine, anti-phosphothreonine and anti-phosphotyrosine.	Proteins modified post-translationally at protein-bound serine, threonine and tyrosine residues.
Phosphorylated epitopes located on specific proteins	Monoclonal antibody Ta51.	Recognizes the neuronal-specific intermediate filament protein neurofilament heavy chain when it is phosphorylated in its 'KSP' domain.
	Anti-phospho-ERK 1/2.	Recognizes the mitogen-activated protein kinase ERK 1/2 (MAPK-ERK1/2) in its activated state due to phosphorylation at a specific site within the epitope.
Cell death markers	Anti-activated (cleaved) caspase.	This antigen shows elevated levels in apoptotic cells.

low in the starting material. Indeed, in the case of antibodies being developed for therapeutic applications, a very high level of purity is required, and the preparation must be free of endotoxins. Purification is typically achieved by affinity chromatography in combination with other chromatographic and non-chromatographic approaches optimized for the antibody of interest (see Chapter 7).

Antibodies are particularly useful as biomedical and bioscience research tools for the detection of or for studying changes in the levels and cellular or tissue distribution of antigens and post-translational modifications, as summarized in Table 9.3. Detection of antibody–antigen binding is achieved by labelling reporter molecules to at least one of the antibodies used in the immunoassay in question. Direct detection of reactivity involves the use of a primary antibody that is labelled with a reporter molecule, whereas indirect detection methods use unlabelled primary antibody, which is subsequently detected by the binding of a labelled secondary antibody. This principle is demonstrated in Figure 9.2, where labels can take a variety of forms including enzymes, fluorophores and small molecules and is discussed in more detail later.

9.3 Immunoassay Methods

The most common immunoassay format is 'solid phase' which means that the antibody or antigen is immobilized on a solid support material. For example, in an enzyme-linked immunosorbent assay (ELISA) procedure, antigen or antibodies are typically immobilized on the plastic surface of the wells of a 96-well microtiter plate. Other examples of solid-phase surfaces are magnetic beads, polysaccharide beads, glass and nitrocellulose.

Most assays use labelled antibodies for detection of antibody–antigen binding. Labelling can be performed on the primary antibody (which was raised to the antigen of interest – direct detection) or on the secondary (i.e. an antibody used to detect the immunoglobulin (Ig) type of the primary antibody – indirect detection). Method selection depends on a variety of criteria such as the affinity and avidity of the primary antibody. Indirect labelling amplifies the signal strength and several labelled secondary antibody molecules

162

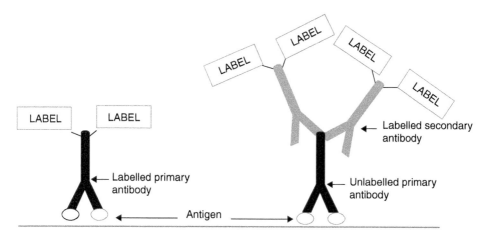

Figure 9.2 *Principles of antibody labelling for direct and indirect detection methods. Shown is a schematic diagram illustrating the use of a labelled primary antibody for direct detection immunoassay methods and the use of a labelled secondary antibody for indirect detection methods. In the latter, the secondary antibody recognizes multiple epitopes in the primary antibody's Fc domain, thus amplifying the signal.*

can bind to the Fc domain of the primary antibody. For example, if the primary antibody were a mouse IgG monoclonal antibody to the antigen of interest, the secondary antibody would be raised in another species (e.g. rabbit) to recognize the whole molecule of the Ig class and species of the primary antibody (e.g. rabbit anti-mouse IgG). Labelling occurs in the Fc domain so as not to interfere with antigen–antibody binding. A variety of labels including fluorophores, radioisotopes, biotin and enzymes can be used with selected examples shown in Table 9.4. Immunoassays can be **competitive** or **non-competitive**. Typical examples of competitive assays are radio-immunoassays and certain types of ELISA.

9.3.1 Competitive Assays

These assays normally use a single-site antigen (i.e. they target an antigen in which the antibody used recognizes a single epitope). Therefore, such assays work well with a single monoclonal antibody, which recognizes one specific epitope. In this type of assay, it is necessary that either the antibody or antigen are limiting. Commonly, such assays are used to quantify the concentration of antigen in a sample. Known amounts of labelled antigen are applied to compete against the binding of unlabelled antigen, the latter being present in the sample or used in the form of a range of concentrations of purified antigen to allow the construction of a standard curve. Labels, which enable the detection of antibody–antigen binding, can either be radioisotopes (e.g. Iodine^{125}) or enzymes (e.g. horseradish peroxidase – HRP).

A schematic diagram of the principles involved in competitive immunoassays is shown in Figure 9.3, which also shows the inverse relationship between the signal (radioactivity or absorbance) and concentration of antigen in the sample. The dynamic range of linearity for such assays is typically 2–3 orders of magnitude.

Key manipulations in competitive immunoassays:

- First of all, a limiting concentration of an antibody raised to the antigen of interest is adsorbed to a solid surface, which in ELISA would be the surface of the wells in a multiwell plate (see Chapter 3). As the capacity of the solid surface is finite, this will limit the amount of antigen that can bind.
- Any unsaturated protein binding sites on the solid surface is then blocked using a non-specific protein (i.e. one that is not recognized by the antibody; e.g. bovine serum albumin (BSA)). This blocks any remaining protein binding sites, thus preventing non-specific binding of antigen to the support material.
- Next, a fixed amount of labelled antigen is added in the presence and absence of different amounts of unlabelled antigen. This could be present in the sample of interest (in which antigen levels are to be

Table 9.4 *Molecules used in antibody labelling to facilitate detection of antigen. Shown are examples of commonly used antibody labels and information about their typical uses and detection methods.*

Label	Immunoassay application(s)	Detection methods	Notes
Enzymes (e.g. horseradish peroxidase [HRP] and alkaline phosphatase)	(i) Immunoblot analysis (ii) Immunohistochemical staining (iii) ELISA	Enzyme-catalyzed substrate conversion into a coloured reaction product.	(i) Typically linked to the oxidation of luminol under oxidising conditions (enhanced chemiluminescence – ECL) in immunoblots. (ii) Insoluble coloured reaction product deposits in vicinity of antigen. Visualized by bright-field microscopy. (iii) Reaction usually stopped by addition of H_2SO_4, giving a measurable yellow colour.
Fluorophores such as fluorescein and tetramethyl rhodamine isothiocyanate (FITC and TRITC)	Immunofluorescence staining of cells and tissue sections. Sometimes used for probing immunoblots.	Fluorescence microscopy for immunofluorescence staining. A fluorescence blot imager is used for detection on probed immunoblots.	Each fluorophore requires irradiation at a specific excitation wavelength and emits fluorescence at a longer (lower energy) emission wavelength. If 2 or more fluorophore-conjugated antibodies are used multiple antigen detection is possible in the same sample.
Biotin	All the above.	Biotin binding probe conjugated to an enzyme or fluorophore.	Probes can be a labelled anti-biotin antibody or a labelled biotin binding protein such as streptavidin. This is typically used as an amplification step but can result in higher background.
Radioisotopes (e.g. I^{125})	Radioimmunoassay.	Antibody-bound radioactivity.	Sometimes used in immunoblotting but HRP-conjugated antibody coupled to ECL is preferred as a non-radioactive alternative.
Colloidal gold	Immunogold electron microscopy.	Gold particles are detected as electron-dense (dark) particles on electron micrographs.	Colloidal gold particles can be a variety of sizes from 5 to 50 nm; hence, double labelling is possible. These are typically conjugated to secondary antibodies. However, other gold-conjugated Fc domain binding proteins, such as protein A and protein G can be used in place of secondary antibodies for certain types and species of immunoglobulin.

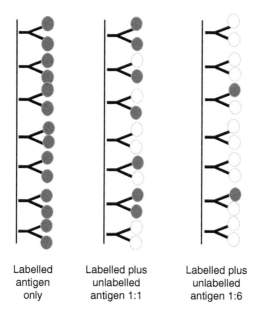

Labelled
antigen
only

Labelled plus
unlabelled
antigen 1:1

Labelled plus
unlabelled
antigen 1:6

Figure 9.3 *Schematic representation of a typical competitive immunoassay. For a radioimmunoassay, the label would be a radioisotope whereas in a colorimetric assay an enzyme such as horseradish peroxidase (HRP) would be used to label the antigen. If increasing amounts of unlabelled antigen are added, the amount of antibody-bound labelled antigen decreases as shown from left to right in panel. This results in a characteristic negative slope as shown in Worked example 9.1. The immobilization of the antibody can be achieved by direct attachment to a surface, as shown here or, more commonly, by reaction with a solid-phase cross-linking reagent that binds the Fc region of the antibody.*

determined) or it could be a series of standards (i.e. pure antigen at known concentrations, to enable the researcher to link signal strength to antigen concentration).

- The plate is incubated for the predetermined optimal time to allow maximum binding of antigen, before the sample is removed and the wells washed with buffer (e.g. PBS containing 0.05% (v/v) Tween 20) to remove unbound antigen.
- The amount of bound antigen can then be measured by the quantity of bound radioactivity (RIA) or by colour development of reaction product, following the addition of substrate (e.g. HRP substrate) for the conjugated enzyme (ELISA).
- The concentration of the antigen can then be calculated. This is inversely proportional to the signal produced by measurement of the labelled antigen bound to the solid phase, which is responsible for the negative slope seen in Worked example 9.1.

This relationship is shown by the following equation:

Concentration of free antigen $\propto$ 1/concentration of solid phase-bound label

This is because the labelled antigen competes with the unlabelled antigen to bind to the Fab region of the limiting concentration of antibody bound to the solid support. Thus, as the concentration of unlabelled antigen increases, it becomes more effective at blocking the binding of labelled antigen, thus reducing the signal strength of the immunoassay. The strength of the signal detected reflects the amount of antigen present in the sample of interest.

- There are certain limitations to competitive assays. For example, assay sensitivity is dependent on the affinity of the antibody for the antigen. Thus, not all antibodies are suitable for this type of assay. Furthermore, the range of linearity (i.e. the dynamic range) between the signal and antigen concentration does not normally exceed 2–3 orders of magnitude. If a sample has a concentration of antigen that is too high (i.e. outside the upper end of the dynamic range), it will only be detectable within the dynamic range after dilution.

Worked Example 9.1 Competitive ELISA

Your task is to determine the concentration of a specific antigen (analyte) in human serum. The average absorbance values from a competitive ELISA for using unlabelled and HRP-labelled purified antigen are given in the table below. The serum sample was diluted 1/20 in phosphate buffered saline (PBS) to obtain absorbance values within the dynamic range of the assay. Calculate the concentration of antigen standards in the serum sample.

Antigen concentration (picomoles ml^{-1})	Absorbance at 405 nm
0.05	0.52
0.1	0.49
0.5	0.321
1	0.242
5	0.093
Sample (1/20)	0.157

To do this task, plot a well-annotated graph using the tabulated data, as shown here using Microsoft Excel. Note that the X-axis in a logarithmic scale.

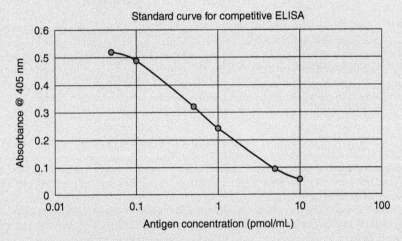

The correlation coefficient is 0.994, indicating a good linear relationship over this range of antigen concentrations (i.e. we have a good standard curve).

The equation for this curve is $Y = -0.093 \ln(X) + 0.2547$.

By transposing the different parts of the equation, we can calculate the natural log (ln) value of the antigen concentration, as follows:

$$0.157 - 0.2547 = -0.093 \ln(X)$$

Assuming Y = absorbance and X = antigen concentration:

$$-0.0977 = -0.093 \ln(X)$$

Therefore, $0.0977 = 0.093 \ln(X)$

$$0.0977 / 0.093 = 1.05 = \ln(X)$$

As this is a natural log value, the antilog must be calculated. This can be done with a good scientific calculator or by searching for an antilog calculator tool online:
The antilog for the natural log value 1.05 = 1.62.
Therefore, the antigen concentration in the diluted sample is 1.62 pmol ml^{-1}.
A quick check of the position where an absorbance value of 0.157 intersects the standard curve confirms that this is a sensible result. If it does not approximate visually to the calculated value, check your calculations again. However, as the sample was diluted 1/20 prior to performing the assay, the original concentration is 20 times higher than the value recorded in the assay.

Final antigen concentration = $1.62 \times 20 = 32.40$ pmol ml^{-1}

Now, try to apply the same analysis to Student Task 9.1.

STUDENT TASK 9.1 COMPETITIVE ELISA

Calculate the final antigen concentration in a sample which, when diluted 1/25, gives an absorbance value of 0.299.

9.3.2 Non-competitive Immunoassays

This type of assay differs from competitive assays in that it can involve not only the use of targeting of a single antigenic site by one monoclonal antibody, but also two sites using 2 monoclonal antibodies or multiple sites using either polyclonal antibodies or several monoclonal antibodies. In such assays, the antibody reagents are present in excess (i.e. they are not limiting), which facilitates a more rapid antigen binding reaction. There are several non-competitive assay formats, some of the most widely used being ELISA (one- and two-site assays), immunoblotting, immunohistochemical staining, immunofluorescence and immunoprecipitation.

9.3.2.1 One-site ELISA

Key manipulations in a one-site ELISA

- Firstly, wells of a 96-well microtiter assay plate are coated with antigen or with sample containing the antigen to be measured.
- Unoccupied protein binding sites are then blocked by incubation (typically for 30–60 minutes) with a protein not recognized by the antibody, such as BSA at 3 mg ml^{-1} in phosphate buffered saline (BSA/PBS). An alternative to BSA is powdered non-fat milk at a concentration of 5% (w/v) in PBS.
- The blocking solution is then removed and replaced with an excess of primary antibody (e.g. a mouse monoclonal antibody IgG) diluted in blocking buffer. Incubation proceeds for a pre-optimized incubation time and temperature.
- Unbound primary antibody is removed by washing in 3 or more changes of PBS.
- The wells are then incubated with enzyme-labelled secondary antibodies which attach to the Fc domain of the primary antibody. Thus, in the case of a mouse monoclonal IgG primary antibody, the enzyme-conjugated secondary antibody must recognize the mouse IgG Fc domain.
- Unbound secondary antibody is then removed by at least 3 washes with PBS.
- Finally, antigen–antibody binding is then determined by adding the conjugated enzyme substrate solution. In the case of a HRP-conjugated secondary antibody, the substrate could be tetramethyl benzidine (TMB) added under oxidising conditions.
- As the reaction proceeds, a coloured reaction product appears (blue). Colour development is typically stopped by the addition of H_2SO_4, and the resultant yellow colour in the case of HRP-conjugated antibodies is measured in a microtiter plate reader (see Chapter 3, Section 3.7.7).
- In contrast to competitive assays, the colour generated in non-competitive assays is directly proportional to the antigen concentration, thus giving a calibration graph with a positive gradient, as shown in Worked example 9.2.

Points to note:

The sequence of key steps in this procedure is: antigen immobilization, blocking, incubation with primary antibody, incubation with labelled secondary antibody and detection of reaction product. The example described above uses the indirect detection method, in which the secondary antibody is labelled (usually by conjugation to an enzyme such as HRP) and the plate wells are coated with antigen. If the primary antibody has high affinity for antigen, assays can also be performed by the direct detection method, in which labelled primary antibody with no secondary antibody step is used. Incubation times and temperatures are optimized for each individual antibody or immunoassay.

Remember that the primary antibody detects the antigen, while the secondary antibody (if used) detects binding of primary antibody to the antigen, acting as a signal amplification step. A summary of some of the most common enzyme labels and their applications is shown in Table 9.4.

Worked Example 9.2 One-Site Non-Competitive ELISA

A study was performed to measure the amount of a specific antigen in human serum using a one-site non-competitive assay with HRP-labelled secondary antibodies. The colour intensity produced after incubation with the HRP substrate TMB was measured in a microtiter plate reader. Use the absorbance values shown in the table below to produce a standard curve for the antigen of interest. The sample was diluted 1/10 to obtain an absorbance value within the dynamic range of the assay. Using the standard curve, calculate the concentration of the antigen in the human serum sample.

Antigen concentration (pmol ml^{-1})	Absorbance @ 405 nm
15	0.071
30	0.221
60	0.429
120	0.691
240	0.991
480	1.328
960	1.599
Sample (1/10)	**0.611**

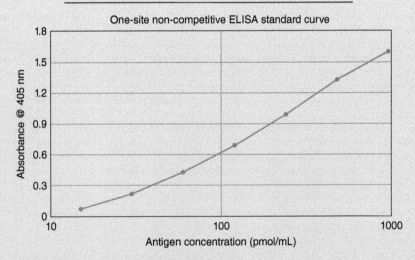

The correlation coefficient R^2 is 0.9873, indicating a good linear relationship over this range of antigen concentrations (i.e. we have a good standard curve). The equation for this curve is Y (given as 0.611) = 0.3793 ln(X) − 1.0541.

By transposing the different parts of the equation, we can calculate the log value of the antigen concentration, as follows:

$$0.611 + 1.0541 = 0.3793 \ln(X)$$

Assuming Y = absorbance, X = antigen concentration, and ln = natural log:

$$1.6651 = 0.3793 \ln(X)$$

$$1.6651 / 0.3793 = 4.3899 = \ln(X)$$

As this is a natural log value, the antilog must be calculated. This can be done with a good scientific calculator or by searching for an antilog calculator tool online.

The antilog for the natural log value 4.3899 = 80.6 (to 1 decimal place).

Therefore, the antigen concentration in the diluted sample is 80.6 pmol ml^{-1}.

A quick check of the position where an absorbance value of 0.611 intersects the standard curve confirms that this is a sensible result. If it does not approximate visually to the calculated value, check your calculations again. However, as the sample was diluted 1/10 prior to performing the assay, the original concentration is 10 times higher than the value recorded in the assay.

Final antigen concentration = 80.6 × 10 = *806* pmol ml^{-1}

Now, try to apply the same analysis to Student Task 9.2.

STUDENT TASK 9.2 NON-COMPETITIVE ELISA

Calculate the final antigen concentration in a sample which, when diluted 1/30, gives an absorbance value of 0.423.

9.3.2.2   Two-site ELISA

This approach is also known as a sandwich ELISA and is depicted schematically in Figure 9.4.

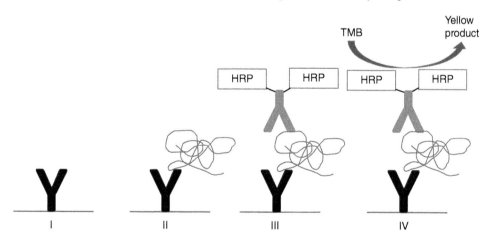

Figure 9.4 *Principles of a two-site ELISA. Schematic representation of a two-site assay. Shown from left to right are (i) the capture antibody attached to solid surface (usually a 96-well plate), (ii) capture of the antigen of interest, (iii) detection of bound antigen with labelled antibody that recognizes a distinct site from the capture antibody (in this case, a form of direct detection using HRP labelling) and (iv) addition of 3,3',5,5'-tetramethylbenzidine (TMB) substrate to detect enzyme label activity and hence the amount of bound antigen.*

Key manipulations in a two-site ELISA

This type of assay requires antibodies to two distinct epitopes that are spatially separated from each other in the antigen structure; one is used to capture the antigen (capture antibody) and the other is used for antigen detection and quantification (revealing antibody).

- An excess of the capture antibody is immobilized onto the solid phase (e.g. microtiter plate well surface in the case of ELISA).
- Unsaturated protein binding sites are then blocked (e.g. using BSA/PBS blocking buffer; see above).
- The sample containing the antigen of interest is then added for an optimized incubation period, having the effect of enriching the concentration of antigen on the well surface.
- Unbound sample is removed by washing with 3 or more changes of PBS.
- The revealing antibody (diluted in blocking buffer), which is labelled (e.g. with HRP) and recognizes a distinct antigenic site to that of the capture antibody, is then added.
- After an optimized period of incubation, unbound revealing antibody is removed by 3 or more washes with PBS.
- The binding of the revealing antibody is then quantified using an appropriate detection system. For example, in the case of a HRP-conjugated revealing antibody, a substrate solution containing TMB would be used as discussed in Sections 9.2 and 9.3, and the resultant yellow colour measured using a microtiter plate reader.

Points to note:

The amount of label detected (i.e. the absorbance values in the example above) is directly proportional to the amount of antigen present in the well. Accurate quantification can be achieved by creating a standard curve using different concentrations of pure antigen in sample wells on the same plate. Assays of this type will typically use two monoclonal antibodies direct at different sites, thus increasing specificity compared to a one-site assay. In addition, the dynamic range of linearity can reach 4–5 orders of magnitude. As reagents are present in excess, the assay is more rapid than a competitive assay and the sensitivity is not usually limited by antibody affinity. This type of assay is very sensitive, being able to detect picomolar levels of antigen.

9.3.2.3 Immunoblotting

Immunoblotting involves the probing of antigens immobilized on either nitrocellulose or PVDF membrane filters. The two main types of immunoblotting are dot-blotting and Western blotting. The former involves the application of a solution containing antigen onto the membrane, whereas the latter first requires the separation of proteins by SDS-PAGE or 2D-PAGE (Chapter 6, Section 2). By using an immunoblotting template device with 96 or more sample wells, dot-blotting can potentially be used for screening large numbers of samples relatively quickly compared to Western blotting. However, it gives no information on antigen size (e.g. protein molecular weight) and is unlikely to detect whether partial proteolysis of the antigen may have occurred. Western blotting on the other hand, while being more time-consuming for high sample throughput, gives more information on the electrophoretic mobility characteristics (size, charge, etc.) of polypeptides containing the targeted epitope. Once immobilized on the protein binding membrane, the key incubation steps are the same for both methods and are summarized below.

Key manipulations in immunoblotting:

- The antigen or antigen-containing sample is bound to nitrocellulose or PVDF membranes.
- Non-saturated protein binding sites on the membrane are blocked, as described above for ELISA.
- The blocked membrane with immobilized sample is probed with unlabelled primary antibody (e.g. mouse monoclonal IgG) diluted in blocking buffer.
- Unbound primary antibody is removed by 3–6 ten min washes in PBS containing a neutral detergent such as Tween-20 (typically at 0.05% w/v).

- Membranes are incubated with labelled secondary antibody directed to the Ig class and species of the primary antibody (e.g. HRP-conjugated rabbit anti-mouse IgG).
- Unbound secondary antibody is removed by 3–6 ten min washes in PBS containing a neutral detergent such as Tween-20 (typically at 0.05% w/v). The sodium chloride salt in the PBS and the Tween-20 detergent help to reduce non-specific charge and hydrophobic interactions, respectively.
- Antibody–antigen reactivity is revealed using an appropriate detection system.

Points to note:
If the primary antibody has a very high affinity for the antigen, a directly labelled (e.g. with HRP) primary antibody could be used. If HRP-conjugated antibodies are used to probe blots, one of the most common detection systems is enhanced chemiluminescence (ECL), which couples HRP activity to the release of light by luciferase-catalyzed oxidation of luminol, which can be detected by special photomultiplier cameras incorporated into blot imaging systems.

9.3.2.4 Dot-blotting

Samples are applied in small volumes and allowed to dry onto the membrane before blocking. It is important to keep the diameter of the dot-blotted material as small as possible to avoid a weaker signal. Using a multiwell manifold apparatus linked to a suction pump will allow for consistent spot size. However, a potential problem with the latter is overflow of sample into adjacent well areas if the manifold is not correctly sealed.

Using a range of sample concentrations allows the determination of the range of linearity between the strength of the detection signal. This would allow the calculation of relative changes in protein levels, provided that signals fall within the dynamic range, which may sometimes require sample dilution. However, if available, a range of pure antigen concentrations can be applied to the same membrane and used for the construction of a standard curve. This would allow a more precise quantification to be achieved. After development, digital images are made in a blot imager and the spot intensities can be quantified using densitometry software, some of which is freely available online (e.g. Image J). A digital image of a probed dot-blot is shown in Figure 9.5.

9.3.2.5 Western Blotting

In this approach, proteins are first separated by gel electrophoresis (see Chapter 6, Section 6.2). The resultant unfixed and unstained gel is carefully placed in contact with a nitrocellulose or PVDF membrane filter and sandwiched between filter papers as shown in Figure 9.6. Separation can be achieved by wet blotting or semi-dry blotting. In the former, the sandwich assembly is clamped into a cassette and fully immersed in continuous transfer buffer (CTB: Typical composition, 20% v/v methanol, 0.01% (w/v) SDS and 192 mM glycine in 25 mM Tris-base, pH 8.3) in an electrophoresis tank similar to that used for gel electrophoresis. In semi-dry blotting, the filter papers placed either side of the gel and membrane are soaked in CTB and the sandwich placed horizontally in the semi-dry blot apparatus. Semi-dry blotting tends to require shorter electrophoresis times and smaller volumes of CTB than does wet blotting. The recent introduction of fast transfer blotters (e.g. Turbo blotters, BioRad, UK) reduces the transfer time to a few minutes. However, it is believed in some research laboratories that wet blotting may be more efficient at transferring certain high-molecular-weight proteins from the gel onto the membrane filter. Whichever method is used, the result is the electrophoretic migration of proteins from the gel on to the membrane, to which they bind irreversibly. This creates a replica of the gel pattern on the membrane filter, as shown in Figure 9.6. As the membrane is thinner than the gel, membrane-bound proteins are more accessible to antibody probing than they would be within the polyacrylamide gel. An image of a probed and developed Western blot is shown in Figure 9.5. Relative differences in the levels of electrophoresed proteins can be determined by densitometry, as explained above for dot-blotting.

Points to note:
In all immunoblot assays, quantification is best performed with the use of an internal control, which could, for example, be a housekeeping protein (i.e. a protein present in all cells), such as glyceraldehyde-3-phosphate

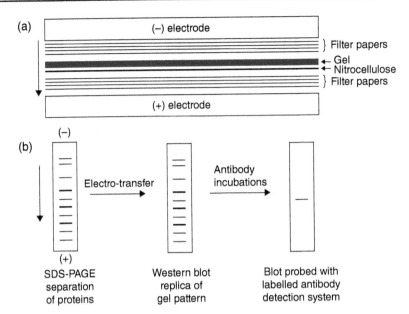

Figure 9.5 *Schematic representation of Western blotting and probing Western blots. Shown are (a) the blotting cassette comprising a sandwich of gel and nitrocellulose membrane filter sandwiched between two sets of filter papers and (b) protein separation in SDS-PAGE, electrophoretic transfer onto nitrocellulose membrane filters and detection of antigen. Arrows indicate direction of protein migration in the Western blot apparatus (a) and in SDS-PAGE (b).*

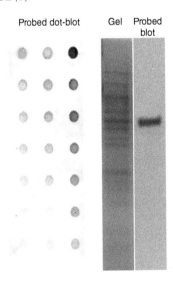

Figure 9.6 *Digital images of a typical dot-blot and Western blot. Shown on the left is a digital image of a dot-blot of 3 different brain microtubule extracts applied onto a nitrocellulose membrane in 1 : 1 serial dilutions from top to bottom. On the right-hand panels are an image of a Coomassie Brilliant Blue-stained SDS-PAGE separation (see Chapter 6) of a brain extract, showing the presence of many proteins, and to its right the corresponding Western blot. In both blots, tubulin (55 kDa) was detected using an anti-tubulin primary antibody and a horseradish peroxidase-conjugated secondary antibody. An enhanced chemiluminescence (ECL) detection system was used to reveal the antibody–antigen binding interactions. Intensity of the reaction is indicative of the amount of antigen present.*

dehydrogenase (GAPDH) or β-actin, which accounts for any variation in the loading of cellular proteins in SDS-PAGE. Changes in the light intensity revealed on development of antibody reactivity with the antigen of interest can be normalized to any changes observed in the housekeeping protein. Alternatively, total protein on the gel can be quantified by staining blots with a reversible protein stain such as Ponceau Red or copper pthalocyanine. Some manufacturers produce 'stain-free' gels containing trichloroethanol which, after electrophoresis, can be activated to visualize separated proteins under UV light. Indirect detection controls should also include a blot incubated with no primary antibody, which would account for any non-specific binding of the secondary antibody (negative control). Positive controls could include a sample enriched in target antigen, or a primary antibody against a different protein but which uses the same secondary antibody as the primary antibody against the antigen of interest.

9.3.2.6 Immunocytochemical and Immunohistochemical Staining

Enzyme-labelled and fluorophore-labelled antibodies are used in immunoassays that reveal the presence and distribution of antigens within cultured cells and tissue sections. If the antigen is cell type specific (i.e. not widely distributed, such as housekeeping and other ubiquitous proteins), its detection can be used to identify that specific cell type. For example, antibodies to choline acetyl transferase will stain only cholinergic neurons in nerve tissue sections and could therefore be used to monitor the abundance and distribution of cholinergic neurons.

The key manipulations include several steps in common with those in the previously mentioned non-competitive immunoassays, except that the cells or tissue sections need to be fixed and 'extracted' prior to blocking, antibody incubations, washes and that the final steps to reveal the antigen location require a light microscope or a fluorescence microscope (see Chapter 2); this is shown schematically in Figure 9.7.

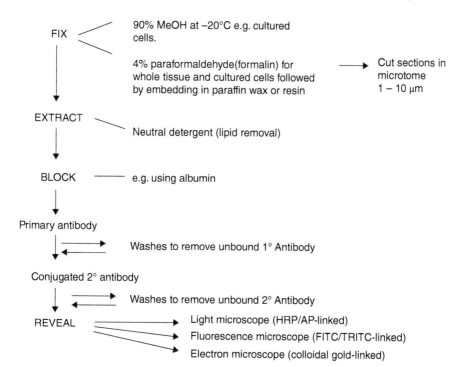

Figure 9.7 *Flow chart depicting key steps involved in the immunostaining of biological samples. In most cases, samples are fixed prior to the extraction, blocking, antibody incubation and washing steps indicated. As indicated, the type of microscope used for visualization of antibody–antigen interaction is dependent on the nature of the label used.*

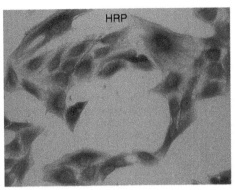
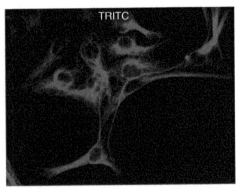
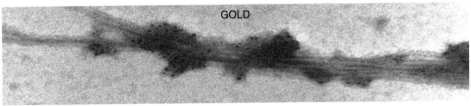

Figure 9.8 *Typical images of immunocytochemical staining. The upper images show cultured glial cells fixed with formalin and stained with anti-tubulin antibodies, followed by secondary antibodies that were conjugated with horseradish peroxidase (HRP) or tetramethyl-rhodamine isothiocyanate (TRITC). HRP staining is detected as a brown deposit (dark in this image), highlighting microtubule networks using a light microscope. Other cell structural detail (e.g. nucleus and plasma membrane) can also be seen using bright-field optics (see Chapter 2). On the right-hand image, only the microtubules are detected as red fluorescence (bright in this image) against a dark background. The lower panel shows a glutaraldehyde-fixed cytoskeletal extract viewed in a transmission electron microscope after probing with a primary antibody to a cytoskeleton-associated antigen, followed by detection using immunogold-labelled secondary antibodies and staining with uranyl acetate. In this case, ultrastructural detail can be visualized (cytoskeletal fibres) and dark spots (colloidal gold particles) are indicative of antibody–antigen binding.*

As shown in Figure 9.8 (upper left panel), when enzyme-linked antibodies are used to detect primary antibody binding to antigen, an insoluble dark-coloured reaction product deposits in the vicinity of the antigen, and the rest of the cell can be seen against a light background. In the case of immunofluorescence staining, which uses fluorophore-conjugated antibodies, the location of the antigen is highlighted against a dark background (Figure 9.8 upper right panel). With the aid of image analysis software, such as Image J, it is possible to determine changes in relative expression of an antigen, particularly if staining is by immunofluorescence. A suitable control for quantification could be immunofluorescence staining of a housekeeping protein such as GAPDH or nuclear staining with a DNA binding dye such as propidium iodide or 4′,6-diamidino-2-phenylindole (DAPI).

If a study is at the ultrastructural level, colloidal gold-conjugated secondary antibodies (or protein A or G) can be used. The presence of the antigen is indicated by the appearance of electron-dense gold particles adjacent to structures containing the antigen (Figure 9.8 bottom panel).

Points to note:
Tissue fixation can sometimes lead to loss of antigen reactivity. Unmasking treatments, such as microwave exposure or protease activity, can sometimes help to reveal the epitope and make it more accessible to primary antibody binding. Formalin fixation can also be a problem if the antigenic epitope contains a lysine residue, as protein bound lysine is chemically cross-linked by this fixative. In such cases, if the antigen is being studied in cultured cells, an alternative fixation procedure (e.g. using 90% (v/v) methanol) may be more suitable.

Non-specific binding can also be a problem with tissue staining, in some cases due to binding with endogenous IgG molecules. For example, if using mouse monoclonal antibodies to detect antigens in mouse tissue, a high background signal can be observed. In such cases, using antibodies raised in another species (e.g. rabbit monoclonal or polyclonal antibodies, if available) might be more effective. A control using an antibody directed to another antigen detected by the same secondary antibody is another useful control to check the efficacy of the secondary antibody.

9.4 Controls

An important consideration in any immunoassay is the use of appropriate controls to confirm antibody specificity. Controls are advisable to check for antibody cross-reactivity with other antigens that may be closely related to the antigen of interest or contain a small region of identical sequence to the epitope. Ideally, both positive and negative controls should be included.

These controls will depend on the antigen of interest and the properties of the antibodies used in the immunoassay in question. Appropriate controls could include:

i. A non-specific immunoglobulin of the same Ig type.
ii. Immunoassay of cells, tissues or extracts that either lack or contain large amounts of the target antigen.
iii. Immunoassay without the primary antibody incubation step (if using indirect labelling).
iv. Normalization controls if quantification is desired (e.g. immunoassay of a housekeeping protein or, in cell/tissue immunoassays, DNA fluorescence staining to account for total cell number in each field of view).

9.5 Summary

- Antibodies are immunoglobulin molecules that recognise specific antigens. There are five classes of antibodies with different functions in the immune system. Antibodies that protect against diseases can be induced by vaccination, but antibodies can also be produced in the laboratory to recognise specific antigens.
- Purified antibodies can be used as therapeutic agents or to identify disease biomarkers in diagnostic tests. They can also be used as molecular tools in a range of immunoassays to identify and/or quantify antigens in cells, tissues, and body fluids.
- Several assay formats exist., which can be competitive or non-competitive in nature.
- Primary or secondary antibodies are typically labelled with reporter molecules (e.g. biotin, HRP or fluorophores) to facilitate detection of antigen antibody binding.
- Appropriate positive and negative controls should be included in immunoassays to account for the possibility of non-specific binding and to verify the condition of the reagents.

SUGGESTIONS FOR FURTHER READING

Most biochemistry and cell biology textbooks will contain some supplementary information on selected laboratory techniques. For example:

- Berg, J.M., Stryer, L., Tymoczko, J.L., and Gatto, G.J. (2019). *Biochemistry*, 9e. WH Freeman. ISBN: 1319114679.
- Nelson, D.L. and Cox, M.M. (2021). *Lehninger Principles of Biochemistry,* 8e. WH Freeman. ISBN: 1319228003.
- Alberts, B., Johnson, A., Lewis, J. et al. (2014). *Molecular Biology of the Cell,* 6e. Garland Science. ISBN: 9780815344322.
- Lodish, H., Berk, A., Kaiser, C.A. et al. (2016). *Molecular Cell Biology*, 8e. WH Freeman and Co. ISBN: 9781464187445.

More detailed information on general laboratory techniques can be found in:

- Hofmann, A. and Clokie, S. (2018). *Wilson and Walker's Principles and Techniques of Practical Biochemistry*, 8e. Cambridge University Press. ISBN: 978-1316614761.

Supplementary Reading for Specific Chapters

CHAPTER 1
Beynon, R.J. and Easterby, J.S. (2003). *Buffer Solutions*. Oxford: Biosis Sci Publishing Ltd. ISBN: 0199634424.

CHAPTER 2
Alberts, B., Hopkin, K., Johnson, A.D. et al. (2019). *Essential Cell Biology*, 5e. WW Norton & Company. ISBN: 0393680398.

Dykstra, M.J. and Reuss, L.E. (2003). *Biological Electron Microscopy: Theory, Techniques and Troubleshooting*, 2e. Springer USA. ISBN: 1461348560.

Slayter, E.M. and Slayter, H.S. (1997). *Light and Electron Microscopy*. Cambridge: Cambridge University Press. ISBN: 0521339480.

Sanderson, J. (2019). *Understanding Light Microscopy*. Wiley. ISBN: 9780470973752.

Thorn, K. (2016). A quick guide to light microscopy in cell biology. *Mol. Biol. Cell.* 27: 219–222.

CHAPTER 3
Gore, M. (2000). *Spectrophotometry and Spectrofluorimetry: A Practical Approach*. Oxford: Oxford University Press. ISBN: 0199638128.

CHAPTER 4
Doerge, R.W. and Bremer, M. (2009). *Statistics at the Bench: A Step by Step Handbook*. New York: Cold Spring Harbor Laboratory Press. ISBN: 9780879698577.

Basic Bioscience Laboratory Techniques: A Pocket Guide, Second Edition. Philip L.R. Bonner and Alan J. Hargreaves.
© 2022 John Wiley & Sons Ltd. Published 2022 by John Wiley & Sons Ltd.

Dytham, C. (2003). *Choosing and Using Statistics. A Biologist's Guide*, 2e. Oxford: Blackwell Science Publishing. ISBN: 1405102438.

Lindley, D.V. and Scott, W.F. (1984). *New Cambridge Statistical Tables*, 2e. Cambridge: Cambridge University Press. ISBN: 9780521484855.

Quinn, G.P. and Keough, M.J. (2002). *Experimental Design and Data Analysis for Biologists*. Cambridge: Cambridge University Press. ISBN: 0851993494.

CHAPTER 5

Adli, D.E.H. (2021). *Separation and Fractionation Techniques: Dialysis-Filtration and Centrifugation*. Our Knowledge Press. ISBN: 6204071114.

Graham, J. (2001). *Biological Centrifugation*. New York: Garland Press. ISBN: 1859960375.

Naidoo, S. (2018). *Centrifugation Techniques*. Burlington, ON: Arcler Press. ISBN: 1773610619.

CHAPTER 6

Brown, T.A. (2000). *Essential Molecular Biology: A Practical Approach*, 2e, vol. 1. Oxford: Oxford University Press. ISBN: 0199636427.

Hames, B.D. (2002). *Gel Electrophoresis of Proteins: A Practical Approach*, 3e. Oxford: Oxford University Press. ISBN: 019963640.

Manchenko, G.P. (2002). *Methods of Detection of Specific Enzymes from: Handbook of Detection of Enzymes on Electrophoretic Gels*. CRC Press. ISBN: 9780367454616.

Michov, B. (2020). *Electrophoretic Gels. CRC Press Electrophoresis-Theory and Practice*. De Gruyter Press. ISBN: 9783110330717.

Nowakowski, A.B., Wobig, W.J., and Petering, D.H. (2014). Native SDS-PAGE: high resolution electrophoretic separation of proteins with retention of native properties including bound metal ion. *Metallomics* 6 (5): 1068–1078.

CHAPTER 7

Bonner, P.L.R. (2018). *Protein Purification (the Basics)*, 2e. Taylor and Francis. ISBN: 1138312479.

Lundanes, E. (2013). *Chromatography: Basic Principles, Sample Preparations and Related Methods*. Wiley-VCH. ISBN: 9783527336203.

Poole, C.F. (2002). *The Essence of Chromatography*. Elsevier Press. ISBN: 0444501991.

CHAPTER 8

Harrison, M.A. and Rae, I.A. (2001). *General Techniques of Cell Culture. Handbooks in Practical Animal Cell Biology*. Cambridge: Cambridge University Press. ISBN: 9780521574969.

Rhodes, P.M. and Stanbury, P.F. (2001). *Applied Microbial Physiology: A Practical Approach*. Oxford: Oxford University Press. ISBN: 0199635781.

Aschner, M. and Costa, L. (2020). *Cell Culture Techniques*. New York: Springer Protocols, Springer. ISBN: 1493992309, 9781493992300.

CHAPTER 9

Hnasko, R. (2016). *ELISA: Methods and Protocols. Volume 1318 of Methods in Molecular Biology*. New York: Springer. ISBN: 1493953850, 9781493953851.

Jefferis, R., Kato, K., and Strohl, W.R. (2021). *Structure and Function of Antibodies*. Basel: MDPI. ISBN: 9783039438976.

Renshaw, S. (ed.) (2017). *Immunohistochemistry and Immunocytochemistry: Essential Methods*. Wiley Blackwell. ISBN: 9781118717776.

Detailed information on laboratory techniques can be usually found on the websites of the manufacturers of laboratory equipment. Examples include:

- http://agilent.com for information on HPLC equipment and other chromatography systems
- http://beckman-coulter.com for information about centrifugation
- http://www.phenomenex.com for information on chromatography columns
- http://www.gehealthcare.com for information on protein purification systems and other bioscience equipment
- http://www.bio-rad.com for information on electrophoresis and other bioscience equipment
- https://www.perkinelmer.com/ for information on HPLC equipment and other chromatography systems
- http://www.waters.com for information on HPLC and other chromatography systems
- www.sigmaaldrich.com for reagents antibodies and cell culture protocols.
- https://www.thermofisher.com for reagents antibodies and cell culture protocols.

More information on statistics and statistical tables can be found at:

- http://www.dur.ac.uk/stat.web/
- http://www.statsoft.com/textbook/distribution-tables/#z
- http://web.abo.fi/fak/mnf/mate/kurser/statistik1/StaTable.pdf

INDEX

Page numbers in *italics* refer to illustrations; those in **bold** refer to tables

Basic Bioscience Laboratory Techniques: A Pocket Guide, Second Edition. Philip L.R. Bonner and Alan J. Hargreaves.
© 2022 John Wiley & Sons Ltd. Published 2022 by John Wiley & Sons Ltd.